TO DERRICK AND HEATHER

ETOSHA

LIFE & DEATH ON AN AFRICAN PLAIN

MITCH & MARGOT REARDON

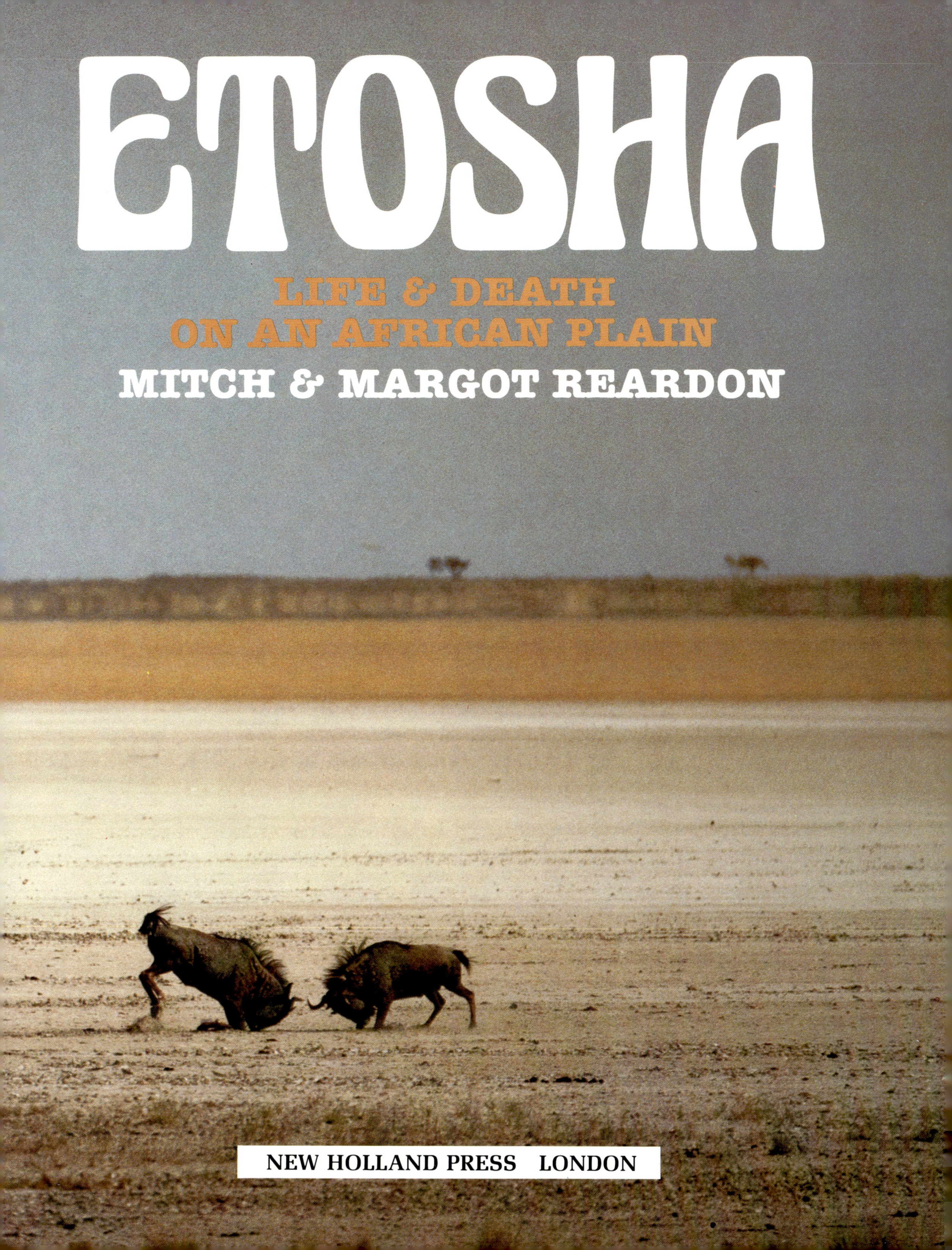

NEW HOLLAND PRESS LONDON

New Holland Press
37 Connaught Street
London W2 2A7
United Kingdom

First published in the United Kingdom 1988

Design by Walther Votteler
Set by McManus Bros (Pty) Ltd
Lithographic reproduction by Unifoto (Pty) Ltd
Printed and bound by Tien Wah Press (Pte) Ltd, Singapore

ISBN 1 85368 031 1

Contents

The extracts from *Out of Africa* by Karen Blixen, and *The Heart of the Hunter* by Laurens van der Post are reproduced by kind permission of Penguin Books, Harmondsworth, Middlesex.
The quotation from *Memories of a Game Ranger* by Harry Wolhuter is reproduced by kind permission of the Wildlife Protection Society of South Africa.

Acknowledgements

Without the co-operation and active assistance of the various officers of the Directorate of Nature Conservation with whom we worked, this book would not have been possible. A special note of thanks to the Director, Mr B. de la Bat, whose enlightened attitudes and constructive criticism kept us going when we felt like giving up. To Senior Conservator, Peter Lind, whose love of the wilderness was so contagious, and in whose company so many of its secrets were revealed to us – a hope that we will work together again. To all our friends back at base camp, Jack du Preez, Kerneels Verwey, Sarel Burger, Errol van Rensburg – and in the field – Ian Hofmeyr, Chris Eyre, Garth and June Owen-Smith and others – whom we were never able to thank sufficiently at the time. To Angela Bartlett of Photo Agencies whose help was invaluable, and to Jumbo Bailey, whose house in Cape Town was a second home to us. We are indebted to wildlife photographers David and Carol Hughes, Des and Jen Bartlett and Clem Haagner, who unstintingly offered advice and encouragement, and to our editors, René Gordon and Peter Schirmer, who were always on hand to help straighten out crooked sentences and recharge us with their boundless enthusiasm.
Etosha, 1981

Introduction

I came to Africa to be amongst the last great companies of game left wandering free in the world today. I came for the peace and the majesty and the adventure. I came because of a grandfather who understood small boys and wild places; who had the talent to fire the imagination with a desire for broader horizons. And I came because time was running out and soon it might all be gone. My choice of Etosha was arbitrary. At first glance it seemed the essence of everything I sought and later, on coming to know it – its communities and conventions – my affection and involvement grew. The world is in trouble, and if the solutions seem obvious, their implementation is less so. Etosha represents a release, a glimpse into the past, a chance to savour a more benign age. It offers no clues and holds out no hope to a frenzied human race driven by their burgeoning populations. It is simply there. The only relevant question is – for how much longer?

Etosha is situated at the heart of Namibia, an insignificant parcel of land crouched on the south-west flank of the African map. The territory is currently engaged in a war that threatens to parallel the recent savage and embittering conflict in nearby Zimbabwe.

Below: *Fleeing one of the many dangers that daily threaten Etosha's creatures, a wildebeest surges across a water-hole.*
Right: *A lone oryx bull faces a charging herd of elephant with the sang-froid of a veteran who has seen it all.*

In spite of its status as a wildlife refuge, Etosha cannot expect to be exempted from the fighting. It lies just south of the Angolan border in what is known as the 'operational area'.

In the end, however, after the last shot has been fired and the soldiers have returned to their barracks, an accounting will have to be taken. The political issues will be decided in a game of winner takes all, and history clearly indicates who the winner is to be. The future of Etosha is far less clear-cut. If it is to become a battle-field, with military priorities superseding the interests of its wildlife, its cause seems hopeless. With so much at stake, the sacrifice of Etosha may seem a small enough price to pay for the fruits of ultimate victory.

Cries of dismay concerning the diminishing wilderness have emerged from Africa for many years now. 'Save our Heritage' and similar appeals have become overworked clichés. Yet they are none the less true for all that. In spite of sympathetic murmurings, far too few proposals become realities, and human needs, in the form of economic development and an expanding population's land hunger, invariably are given precedence. A handful of determined individuals and organizations desperately seek to reverse the trend. These range from the research worker in the field, who is just beginning to understand what makes our environment what it is, to the newspaper columnist who clarifies the issues so that none can say 'we didn't know'.

I have no wish to harp on the inherent dangers. Rather, I would sound a warning now, at the beginning, and go on to show Etosha as it revealed itself to me in the three years I lived there. What I saw and experienced can only partially be portrayed in print. No photograph or word can fully convey the stillness of that ancient land, its hugeness and fragility. Contradictions and anomalies abound. Life and death, and their interaction, are the only constants. The experience is awesome and humbling.

Mitch Reardon, Okaukuejo, June 1981

Out on a hunt, this female cheetah and her four cubs sit sphinx-like on the crest of a hillock, watching for prey.

Communities

There is a lake in Africa that died millions of years ago. It died of thirst during an arid era in the earth's history when the rivers that fed it ceased to flow. Gradually evaporation and seepage dried it out, leaving only its skeleton – a cracked, white salt pan over six thousand square kilometres in extent. But the bones of the old lake are still possessed by its spirit and, in years of good rains, water collects in this 'place of mirages' to display it as it might have looked all those millennia ago.

The pan still drains the water from the north, though the flow is but a mere fraction of its former volume. During the wet season run-off rain-water meanders down a network of shallow watercourses to converge on Lake Oponono, north of Etosha in adjacent Ovamboland. From here the Ekuma and Oshigambo rivers, the last legs of the southern journey, debouch the water into the Etosha Pan. Another artery, the Omuramba Ovambo, receives its water from a catchment area to the north-east of Etosha and enters through a delta known as Fischer's Pan.

In living memory, Etosha has never been full of water, but, in years when good rainfall causes partial flooding, it holds large sheets of shallow water, usually not more than a metre deep, and in places only a few centimetres. The effect created by these ephemeral waters is of a tranquil lake spreading away to the horizon. Thousands of waders and other waterbirds, gathered along the beach or probing the shallows, enhance the image and lend an air of permanency to a phenomenon that at best maintains itself for only a few short months.

Beyond the pale calcrete of the pan's southern shore, a haze of lion-coloured plains merges into dense, anonymous mopane and acacia wood-land. Both the grasslands and the forests support a complexity of animal, bird and insect life. It is exciting, alien country; immense, inscrutable.

'It was the Africa I had read of in books of travel,' exulted the American trader Gerald McKiernan in 1876. 'All the menageries in the world turned loose would not compare to the sight I saw that day.'

To reach the pan, McKiernan had crossed a stretch of country then so rugged, so lacking in surface water, as to be almost inaccessible. He had laboured with his ox-wagon from the Atlantic coast, through the hostile Namib desert and on into a broken wilderness of thorn scrub sparsely inhabited by sullen tribes. It had been a brutal, unremitting struggle, and his delight in his new surroundings is infectious. Before him lay a rolling grassland, richly stocked with all manner of wild beasts, and immediately beyond lay the kingdom of the Ovambos where he hoped to establish a lucrative trade in cattle and ivory.

'Travelled all day through a beautiful country', he continued, 'almost a level plain dotted with clumps of timber, in the middle of which pools of water left by rains were to be found, some of them of considerable extent. Wild ducks and geese were plentiful, the latter of beautiful plumage and with crested heads. It was by far the prettiest scene I had met with, and I found it difficult to realize that I was in desert Africa. That afternoon we fell in with immense numbers of animals beyond anything I had yet seen. I

Previous page: *In the quiet of a grey sunrise, fording zebra ruffle a quiet Fischer's Pan.*
Right: *Reacting instinctively to a charging lion, zebras explode from a water-hole. They respond to any threat in instant, panicked flight, though should the alarm prove false they quickly settle down again. But they act first, and think later, analysing danger while already on the run.*
Far right: *Sparked by her presence, a pair of stallions lunge, bite and kick in competition for the favours of a mare in oestrus. Phases in such ritual clashes among males from different groups intensify until, if neither gives way, they reach this climax.*

would scarcely be believed, if I should state that there were thousands of them to be seen at a sight. Gnus in herds like buffalo on the plains, hundreds of zebras, beautiful in their striped coats, springboks by tens of thousands, ostriches, gemsboks and steenboks, hartebeeste and elands. Water and grass was plentiful and they seemed to be having an easy time of it, but we soon disturbed their serenity, and that night we had a variety of meats such as no other country can afford: zebra steaks, gnu steaks and broiled ostrich wings.'

Twenty-five years earlier a Swedish trader-explorer, Charles John Andersson accompanied by an Englishman, Francis Galton, had been the first Europeans known to penetrate this remote region.

In his journal, the first publication to describe the area to the outside world, Andersson recalls a visit to Omutjamatunda, as the first Ovambo cattle post at Namutoni was then called – 'there is a most copious fountain, situated on some rising ground and commanding a splendid prospect of the surrounding country. It was a refreshing sight to stand on the borders of the fountain, which was luxuriously overgrown with towering reeds and sweep with the eye the extensive plain encircling the base of the hill; frequented as it was, not only by vast herds of domesticated cattle, but with the lively springboks and troops of striped zebras. If the monotony of our dreary wanderings had not thus occasionally been relieved, I do not know how we should have borne up against our constant trials and difficulties.'

If the early explorers found relief and a sense of well-being in the shadow of Etosha, so did a remarkable diversity of wild creatures. Permanent springs and nutritious grasslands attracted the plains game – herds of zebra, wildebeest and springbok, while in the adjoining tangle of acacia and mopane bushveld, browsers such as kudu, giraffe and the reclusive rhino thrived. With such a wealth of herbivores, the predators were also well represented.

Moving to natural rhythms dictated by the availability of food and water, the animals prospered in their austerely benign environment. The only resident humans were small groups of Heikum Bushmen, stone-age hunter-gatherers living in harmony with their surroundings.

Etosha today is not the wilderness it was then. Good roads and tourist lodges have softened the impact of nature in the raw. Yet sitting alone, watching a cheetah saunter across a plain, a plover starting up before it, screeching in alarm, or ground squirrels nervously seeking shelter at the approach of a foraging jackal, it is easy to imagine that little has changed.

At times like these, sights and sounds were indelibly imprinted on my consciousness. Images of Africa, of two giraffe, still as killed trees, silhouetted against an ancient sky; the light in the eye of a fugitive fox; and the insistent cry of a solitary crow reiterating some annoyance a hundred times.

One of my most vivid first impressions of Etosha was the spectacle of more than a thousand zebra, together with hundreds of wildebeest and springbok, plodding in single file in a daily mini-migration from their pastures at Poacher's Point to the Andoni water-hole. Two parallel columns reached across the flat savanna from the horizon to the water, one approaching to drink and the other returning. The wild, rallying bark of zebras and the mournful bawling of wildebeest complemented the strangely Pleistocene tableau. At Andoni the milling crush was enormous, the animals alternatively wading in to drink, standing shoulder to shoulder, until an individual would snort at some imagined danger, causing all to stampede in a flurry of white water.

Left: *Out on the plains an excited springbok stots in a series of stiff-legged bouncing leaps – a characteristic action which expresses either exuberance or alarm.*

Right: *Startled, a black-faced impala takes off in a gravity-defying leap that may be 3 m high and span 10 m. Seconds later the whole herd fled, reacting to her warning.*

Zebra, wildebeest and springbok are essentially short-grass feeders, avoiding thickets and other dense vegetation, and – as the most populous species in Etosha – are easily observed. Their winter domicile is the grassy plains around the southern edge of the pan. During the rainy season, most of them migrate more than 150 kilometres to the nutritious sweetgrass savannas to the west of the pan. They probably select the plains as a measure of protection from predators, as well as for the higher nutritional value of the grass.

Although all three are grazers and frequently feed in close proximity, under ideal grazing conditions there is very little competition between them. A fascinating system of grazing succession has evolved in which each species feeds on a different level of the herb layer, according to its needs and particular way of obtaining protein.

The zebra's digestive system is adapted to gain maximum nutrition from the coarse grass it eats, and at the same time, process a large quantity of fodder. Zebras do not mind grass with a moderately low ratio of leaf to stem. After grazing they move on, leaving the lower sections – more leaf and less stem – for the wildebeest, which need the higher protein content and easier digestibility of this part of the plant. The wildebeest leaves the

high-protein growing shoots and herbs for the springbok, which has the added advantage of being a mixed feeder. When grass is in short demand, it browses the winter-sprouting leaves of various acacias or the protein-high, desiccated twigs of salt bushes.

As the most numerous of the herbivores, zebra, wildebeest and springbok naturally attract the most attention of the various large predators, and as a result have developed techniques designed to reduce the chance of being eaten.

As the preferred prey of lion in Etosha, zebra and wildebeest rely to a great extent on the open nature of their habitat to spot a stalking lion before it can attack. A snort or bark of alarm alerts everything on the plains to the danger, and staring pin-points the exact locality. A lion invariably aborts a hunt on being discovered, sometimes walking off disgustedly in full view. The tension goes out of the herds when a lion so reveals itself, and after it has passed from sight into a thicket, they quickly forget about it and return to their grazing – thus affording the lion another opportunity of taking them unawares.

The dependence of zebra and, particularly, wildebeest on an open terrain to maintain a balanced prey-predator relationship was recently highlighted in the Kruger National Park. Bush encroachment and several seasons of particularly heavy rains led to the wildebeests' traditional pastures becoming overgrown. Not equipped to graze the high-standing grasses, they were forced to concentrate on short-grass 'islands' surrounded by thick vegetation, which offered excellent cover to hunting lions. The consequent slaughter and reduction of the wildebeest population prompted the park authorities to consider the drastic counter-measure of culling large numbers of lion to relieve the pressure. A carefully-monitored experimental cull proved this unfeasible and, ultimately, it could only be hoped that the wildebeest would hold their own until an expected drier climatic cycle re-created more favourable living conditions.

The prey species' most automatic response to danger is flight, but their daily habits also incorporate elementary, yet vital, protective measures such as walking in single file, drinking by day and avoiding dense cover. When attacked they scatter and leap, confusing the predator which must select an individual from a kaleidoscope of swirling, patterned animals. The moment before a lion seizes a zebra from behind, the desperate animal lets loose a powerful backward kick. This kick is so instinctive that even foals who have no hope of benefiting from it lash out. In the case of an adult

This hungry lioness streaked from cover behind a stand of sedge, where she had hidden, unnoticed by a herd of zebra coming to drink. Only when they were within her range did she charge the herd, in an attempt to select an individual from the fleeing kaleidoscope of bodies (Top left). *Closing on a mare that had broken from the massed tumult of stripes and dust* (Centre left), *the lioness was able to seize the zebra's rump, but the mare lashed out with a powerful, instinctive, backward kick that allowed her to escape – and left the lioness rolling in the dust* (Bottom left).
Top: *A lion revealed is a lion disarmed. On the Pan this herd of grazing springbok, unafraid, yet curious and cautious, opens a path for a passing, and disinterested, lioness.*

Left: *The leopard is a creature of darkness – to meet one abroad in daylight is unusual.*
Right: *Clear against the sky, this hunting cheetah scans Halali's plains from her vantage point. Had she not been silhouetted, it is unlikely that I would have encountered her.*

zebra, however, the lion must anticipate the kick, attempt to avoid it and go on to secure a claw-hold on the victim's hindquarter and drag it down.

I have seen a zebra leave an inexperienced young male lion cartwheeling in the dust behind it from an effective kick. In this instance the lion went off chastened but unharmed, although broken jaws and other, sometimes-disabling, injuries are occasionally sustained. Judging by the number of zebras displaying claw-marks and other lion-inflicted wounds, a high proportion of adult members in any zebra herd have evidently experienced the terror of a lion's clutch, and lived.

Because springbok are so numerous, lions sometimes pursue them, but usually they are disregarded, being too swift and too small to justify the effort required. When a lion feeds off springbok venison, it is invariably the result of an opportunist kill or one scavenged from another predator. Springbok, however, form the bulk of a cheetah's kills and a proportion of a leopard's.

Little is known of the habits and prey preference of the secretive Etosha leopard. Throughout its range in Africa, it is recognised as a catholic feeder, preying on those species most readily available. The only leopard kill I witnessed in Etosha involved a springbok that was taken out of a herd moving through mopane woodland. I also watched a leopard try to ambush a guinea-fowl without success. Wildlife cinematographer, David Hughes, filmed two failed hunts – one a reversal of roles in which an irate warthog chased off a young leopard, the other involving the same leopard, where the intended victim, a steenbok, escaped.

Etosha's landscape features abrupt transitions from grassy plains to mixed woodland, as the soil changes. These sudden vegetation shifts cater to natural communities whose needs are best served by each, but there is significant commuting between the habitats. Giraffe which by day browse woodland leaves, move out to the safety of the plains at sunset, to ruminate

and keep a high-rise eye out for lions. Spotted hyenas are at home in either environment, although their shaggy cousin, the brown hyena, is essentially a creature of wide open spaces. Arboreal bush squirrels prosper in the forests, whilst ground squirrel burrows pepper the plains. On the plains large herds are the norm, whereas forest dwellers tend to be more solitary.

The black rhino is a native of the forests and thickets, and this, coupled with its retiring nature, probably accounts for its continued existence in Namibia long after the plains-dwelling white, or square-lipped, rhino was exterminated. The status of the rhino before the white man arrived, and its subsequent fate, becomes clear from the writings of early explorers and sportsmen.

'He is a swinish, cross-gained, ill-favoured, wallowing brute', reported Cornwallis Harris in the early 19th century, 'with a hide like a rasp, an impudent cock of the chin, a roguish leer from out of the corner of his eye, a mud-begrimed exterior, and a necklace of ticks and horse-flies.

'Whether from a limited sphere of vision, arising from the extraordinary minuteness of the eyes, which, resembling a pig's in expression, are placed nearer to the nose than in most other animals – or whether from an overweening confidence in its own powers – the Rhinoceros will generally suffer itself to be approached within even a few yards before condescending to take the smallest heed of the foe who is diligently plotting its destruction. At length, pricking its pointed ears at some unusual sound, it listens with a ludicrous assumption of shrewdness – its elevated snout armed with a double ploughshare, importing an inimitable expression of contempt. In an instant the dull and vacant physiognomy has become lighted up with the essence of all that is spiteful and malevolent. Twinkling its hoggish eyes, and turning its shapeless head inquiringly from side to side – it trots forward a few paces with the vivacity and mincing gait of a French dancing master – wheeling presently to the right about to reconnoitre the enemy. Then uttering a great blast or snort of defiance, and lowering its armed muzzle almost to the ground, grunting and trumpeting, on comes the villain with reckless impetuosity; displaying a degree of activity but ill according with such unwieldly proportions.'

In those early days there seems to have been no shortage of rhino for any sportsman who wished to shoot them – and all of them did. Harris notes that 'gregarious in fives and sixes, they are extremely abundant in the interior, and I have, during a single day, counted upwards of sixty.'

'The chase of the rhinoceros', wrote Charles Andersson, commenting on the ease and relative safety with which the large and fearsome-looking beasts could be dispatched, 'is variously conducted in Southern Africa. One of the most approved plans is to stalk the animal, either when feeding or reposing. If the sportsman keeps well under the wind, and there be the least cover, he has no difficulty in approaching the beast within easy range, when, if the ball be well directed, the prey is usually killed on the spot. With a little precaution this kind of sport may be conducted without greatly endangering a person's safety.

Above: *This diminutive dik-dik – smallest of Etosha's antelopes and no bigger than a lamb – thrives in the eastern woodlands which provide browse and cover. Dik-dik, whose hooves have adapted to the rocky limestone ground, prefer the low rainfall areas of these parts.*
Right: *The shy, largely nocturnal, black rhino is rarely seen by visitors to Etosha though the rhino population is increasing. This nursing cow, disturbed in her noonday siesta near Otjovasandu, peers dimly at the source of her awakening. Her young calf, born after a 15-month gestation period, will remain with her for four to five years, until she again gives birth.*

The yellow, puff-ball flower clusters of the Acacia nebrownii (Centre right) *are in startling contrast to the sere landscape as they burst into bloom during the arid winter months when most of the other trees and shrubs have lost their leaves. Though most of Etosha's browsers eagerly seek the flowers, there is no rigid demarcation of feeding levels and very little competition takes place between the various species. This little steenbok* (Bottom right) *stretches to harvest the lowest blossoms, kudu* (Far right) *concentrate on the middle branches, while the giraffe* (Top right) *feeds off the top of the shrub. Bushfire control has allowed acacia scrub to encroach on the grasslands, extending the browsers' range and encouraging their increase to the detriment of the open savanna grazers such as zebra and wildebeest.*

'The number of rhinoceroses destroyed annually in Southern Africa is very considerable. Of this some idea may be formed, when I mention that Messrs Oswell and Vardon killed in one year, no less than eighty-nine of these animals; in my present journey, I myself shot, single-handed, nearly two-thirds of this amount.'

Most of these rhino hunts were at night, sitting up over water-holes which often constituted the only available water for many kilometres around. As rhino will drink every other evening or even every 24 hours, hunters needed only to sit up for a few consecutive nights to wipe out almost the entire rhino population of an entire area. The temptation was too great to resist, and they set about doing just that. Not long after Andersson's epic hunt, the heyday of rhino shooting had passed. Later sportsmen gloomily reported on the abundance of rhino skeletons without anywhere finding a single living beast.

Rhino, it turned out, were not only easy to shoot, but delicious to eat as well. 'I like rhino flesh more than any other wild animal', wrote Francis Galton, Andersson's fellow traveller. 'A young calf, rolled up in a piece of spare hide, and baked in the earth is excellent. I hardly know which part of the little animal is the best, the skin or the flesh.'

Only in the isolated and mountainous belt dividing the Namib desert from the plains of the interior plateau did a few rhino survive. In 1966 only 90 rhino remained in all Namibia, of which only 15 were protected in Etosha National Park.

To safeguard the rhino's future, the Conservation Department decided to capture and transfer as many as possible to suitable habitats in Etosha. By the end of a complex operation, 43 rhino had been transferred to Etosha and now constitute a viable population. We may never find out what a rhino calf tastes like, but it isn't too late to see what one looks like.

During the vast, but gradual, physical and climatic changes which marked the prehistory of southern Africa, its animals had to adapt to the changing environment – or die. And no creature made this transition of habitat more smoothly than the antelope.

It is thought to have originated in the forests of the Miocene epoch and, as the forests dwindled in the Pliocene and Pleistocene periods, to have successfully occupied the new environments these brought. It acquired great speed, endurance and agility, and grew in size, changing the colour of its coat to fit the pattern of its surroundings. Antelopes came to outnumber all other large mammals and today exist in an amazing diversity of sizes and shapes, coat colours and patterns, and horn configurations.

Of more than 70 different species of African antelope, 15 occur in Etosha. They range from the tiny dik-dik, with a mass of only five kilograms and which stands no more than 400 millimetres at the shoulder, to the world's largest antelope, the eland, up to 850 kilograms in mass and measuring nearly two metres. They include such introduced species as sable and roan antelope, black-faced impala and tsessebe, as well as rare reedbuck unsuited to Etosha's habitats. Some, like the kudu bull with its handsomely

white-barred body, well-bred head and spiral horns, are beautiful. Others like the wildebeest, which possesses the head of a Spanish fighting bull and the body of an ugly pony, are little short of ridiculous. Great companies of springbok are everywhere on the plains, while the nocturnal grey duiker is hardly ever seen. All are herbivores, but some are browsers, eating the leaves of trees and shrubs, creepers and herbs; others are grazers, cropping only suitable grasses; and some select from the best of both worlds according to the season.

The establishment of Etosha as a national park protected the wildlife, but brought new problems. As the herds became more sedentary, they were exposed to infectious diseases, such as anthrax, which tend to build up at certain incubation points. The control of bush fires has been detrimental to the open grasslands which need regular burns to deter invasion by shrubs and tree scrub, while existing scrublands develop into dense thickets. These factors have tipped the scales against the plains game by reducing available pastures and leaving them more vulnerable to predation by lion, which require cover for successful hunting.

There has been a devastating population crash of wildebeest, though their numbers seem to have stabilized; the eland population in eastern Etosha has decreased considerably, though they appear to have increased in the western parts. Predation by lion seems the most important single factor influencing this state of affairs. The lion population in eastern Etosha is increasing apparently because a general thickening of scrubby vegetation around water points has provided better cover for hunting. The large number of carcasses as a result of anthrax during the wet season helps sustain lions at a time when hunting is difficult and a high cub mortality can be expected.

Anthrax is absent from the Otjovasandu area to the west, however, and the relative scarcity of lions there is further attributed to their leaving Etosha for the easy pickings on neighbouring cattle ranches where they are shot by irate farmers.

Not all antelope species have been adversely affected by these disruptions. Such encroachment has favoured kudu and impala. Giraffe, too, have increased and extended their range.

Visitors to game parks are often inclined to resort to anthropomorphic labels when describing the various wild animals they encounter, and a tendency to characterize animals as either good or bad, relative to human values, persists. As a consequence, two of the least loved and most misunderstood animals are the jackal and the spotted hyena.

Intensive research carried out in East Africa on hyenas has revealed many hitherto unknown aspects concerning these much maligned and despised animals. It is now accepted that they are not only scavengers but also powerful and resourceful hunters in their own right. Nor are they incorrigible cowards. In their element, at night, they display a boldness only marginally tempered by pragmatic caution. They have a superb sense of smell and hearing, highly efficient night vision and a functional build. But for all that they are burdened with a poor public image.

Top left: *Shoulder to shoulder, their legs spread for better support, a pair of giraffe bulls battle it out, swinging their massive heads to land sledgehammer blows on the legs, neck and body of their opponent. 'Necking' is not a form of love-play between a male and female as is commonly supposed. Males demonstrate natural affection by languorously entwining their necks, and aggression by stepping-up this action into a vigorous buffeting.*
Bottom left: *Dust flies and a crowned plover flares its wings in protest at a zebra's rolling pleasure. Zebras enjoy their dust baths and have favourite wallows which they visit regularly, and where whole herds line up to take their turn.*

Below: *Moments after confronting each other at the woodland water-hole they both share, two kudu herd bulls clash in a jarring test of strength. Their duel illustrates a fascinating example of 'tail talk'; the dominant bull confidently flashes the white undertufts of his tail, while his losing opponent ignominiously tucks his between his legs.*
Right: *Out in the middle of Halali's treeless plains, a mating pair of bat-eared foxes shelter from the midday heat in the meagre shade of a salt bush.*

People complain that they are loathsome to behold. Hyenas appear misshapen, four-legged Quasimodos. They shuffle along; they lurk. They are the embodiment of all our childhood nightmares of demons and cruel ogres. Their voice is a graveyard moan; they work by night. They gobble down living flesh and are without mercy. They are scruffy, spotted ruffians, slouching about, always looking for trouble.

So, in spite of the writings of naturalists who have come to admire them, hyenas remain beyond the pale of human acceptance. Yet without them and other scavengers, the woodlands and plains would soon become a charnel house. The hyena is well adapted to its role as a scavenger. Amongst its most effective features are its enormously strong teeth and jaws which allow it to chew and digest large bones and tough hide when little else remains of a carcass.

It was thought that hyenas and jackals occupied the same ecological niche, but this has recently been exposed as another example of how far from the real state of affairs the popular image can be. Jackals do not have the specialized jaws and teeth to splinter bones, nor the powerful digestive system that enables a hyena to extract nutriment from bones and offal. Probably no more than 20 to 35 per cent of a jackal's diet is made up of carrion. They also eat insects, as well as rodents, small reptiles, birds and fruit. They are major predators of all but the largest of antelope fawns, and a pack of jackals was reported to have been seen killing an adult springbok ram, which was possibly injured or ill. With few exceptions, carnivores cannot neatly be classified as hunters or scavengers. Given the opportunity or the right set of circumstances, they will fill either role.

Outside of game parks, jackals are classified as vermin. They have become notorious as sheep and poultry thieves. Farmers resort to poison, traps, guns and dogs in an effort to eradicate them. But jackals are extremely adaptable and versatile, each generation becoming more adept at refining its survival techniques.

On one occasion when travelling outside Etosha, I stopped to stretch my legs alongside a road that led through a farming community. The game herds had long since been eradicated from the land, and the soil ploughed over. Then near by, unexpectedly, came the short sharp bark of a fugitive jackal. In that redesigned landscape, the cry was a poignant, ringing echo from the past. Although probably all too few would mourn its passing, the African forests and plains would be infinitely poorer without the call of a jackal in the night.

Far left: *Because of his innate stubbornness this rhino had refused to give way in a water-hole confrontation with an elephant – in spite of his opponent's superior strength. The following morning a nomadic lion had come across the carcass and had begun the slow task of licking and chewing through tough hide to reach the flesh. Lions are opportunistic feeders; they are as likely to catch, scavenge or scrounge their meals as the hyenas they were once thought to support with their left-overs.*
Below: *A circle of griffon vultures watches impassively as two black-backed jackals contest a morsel from a fast-dwindling zebra carcass. Vultures and jackals feed throughout the heat of the day, long after hyenas have quit the scene of the night's kill and departed for their daytime retreats.*
Top left: *At the bitter end of the feast, a pied crow defies a white-backed vulture in a quarrel over the scant remains.*
Centre left: *Revealed in the first light of early morning, a pack of spotted hyenas feed on a kudu bull they had run down in a nocturnal chase. Far from being the mobile garbage disposal units they were once popularly thought, recent research has proved that hyenas are highly efficient hunters, working in co-ordination to make their kill.*
Overleaf: *As day turns to night, a herd of wildebeest move to their evening pastures, alert and aware that it is now the big cats rouse themselves for the hunt.*

The Dry Season

In a dry land people are by nature weather watchers. Whether the rains are good, or late, or fail altogether, is by necessity more than just of passing interest, for it can mean the difference between life and death. And so, towards the end of each dry winter, we consult among ourselves; testing, analysing the omens, prophesying the prospects for summer. Eyes screwed up against the glare, we stare at the enigmatic empty blue sky, but find no answer there. It is a restless and uneasy time.

For a visitor arriving in winter, the bleakness of the country must be almost repellent. At first appearance it is implacably hard and relentless. The parched grasses and bitter thorn, the hard-baked earth, the old sky shrouded by dust, filtering through which a dull sun looms like a blind eye. Yet there is something very reassuring about a majestic, unvarying order to things. The land is tough, lean and disciplined. It gives, and asks, no quarter, and only the best survive.

The advance of the dry season is gradual. It is a slow, almost imperceptible yet constant drying out and dying off. One day the green grass is yellow and still nutritious, the next it is blonde and less so. Then it is gone altogether. Months of austerity and privation lie ahead.

October is the madness month. Temperatures rise and resistance is lowered. The air hangs hot and heavy and still, until a rising thermal boils up to whip dust devils across the desiccated land. The game herds bow, hunched and submissive.

Etosha has no perennial rivers or watercourses, and although temporary rainwater pans are frequented by wildlife during the wet season, soon after the last rains have fallen the pans evaporate and the animals must return to permanent springs. It is in the dry season that these springs come into their own, with all activity centred on the life-sustaining pools. They have perpetuated the plentiful numbers and diverse species populating the Etosha area for thousands of years.

In a daily ritual, zebra and wildebeest move in head-nodding, nose-to-tail columns from their sparse pastures to the receding water levels where they stand massed together sipping the saline moisture at the surface.

Cackling with agitation, blue-jowled guinea-fowl scuttle over calcrete boulders, jostling for position amongst the larger herbivores. A slender mongoose approaching the water is noticed by a pair of blacksmith plovers who chase it, dive-bombing and castigating furiously, from cover to cover. The mere sight of this inveterate egg thief and devourer of fledglings drives all the smaller birds into paroxysms of rage. But moving with a sinuous grace, he eludes them. He is accustomed to such abusive reception.

Fifty metres distant, a knot of kudu stand speculative, appraising their intended route of approach. This is no time for impatience. Then, reassured by the presence of earlier arrivals, they move in – still wary, high stepping, bunched muscles tensed for instant flight. From the shelter of a raisin bush, a fork-tailed drongo watches their tentative advance, chirping and fidgeting, calling the odds with the brashness of a gambler come to try his luck in a strange town.

The dry season is a time of waiting. The parched earth gapes with thirst and the pale sky holds little promise. At this time of the year, the pan is a hostile wasteland of salt-crusted clay, scarred by game trails that meander towards empty horizons.

Top right: *A black-backed jackal seeks relief from a sun that roars overhead, but the ease is temporary for the rain-water pan will soon evaporate.*
Centre right: *In an effort to beat the heat a ground squirrel flares the portable umbrella of its tail to create its own shade. As further protection from the sun it has also scraped loose sand over itself.*
Bottom right: *Springbok line up to drink from a rock fissure that has protected the last of summer's rain from evaporation by the hot sun and winter's dry, easterly winds.*
Far top right: *Mudpacks, which help insulate them against the sun's rays, enhance the already bizarre appearance of these two wildebeest. These animals also have a remarkable physiological adaptation that helps them cope with heat; during the day they can absorb heat so that their body temperature increases considerably. At night the heat is radiated and their temperature returns to normal.*
Far bottom right: *Throughout the dry season, family groups of zebra daily join in large herds to plod in a head-nodding, nose-to-tail mini migration along well-worn game trails to the nearest water-hole.*

All along the edge of the pan, contact springs arise where water flows between the surface limestone and underlying layers of clay. As a rule these springs yield little water, but they are relied on by the herds of gemsbok and wildebeest that graze the salt-loving annual grasses growing on the pan in winter. Each day they come in across the baked clay bed in the heat of late morning, mirage-distorted, swimming above the horizon like images in a surrealist painting. They drink and then depart, the tortured air elongating them into unfamiliar shapes until they appear above the liquid edge of the horizon, linger, then vanish off the edge of the world.

The herds cannot desert the few permanent springs, but because of overgrazing in the immediate vicinity, each day are forced to move further afield in search of fresh pastures. Their dependence on the remaining springs leaves them vulnerable to predators waiting in ambush – and the hunting is good.

Lions loll replete in the shade near convenient pools, stomachs straining against massive loads of meat. Food and heat combine to reduce them to an almost comatose state. As infrequently as possible they change their position to relieve the pressure on bulging bellies and to realign themselves with the shifting shade. As they pant, webs of saliva flow freely from their jaws to settle and clot between their paws. Their good fortune is in direct contrast to the hard times faced by their prey. But when the rains come the situation will be reversed. With abundant water and grazing, the herds will disperse and the lions will be hard put to make ends meet.

The rainfall in Etosha is seasonal; even so its occurrence and distribution is erratic. One area will benefit from soaking downpours while another, not too far distant, is passed over. Such arbitrary climatic behaviour may result in a pocket of drought, misery and death in a land of plenty. Prosperity and misfortune depend upon the rains. Whether they fall or not means the difference between a dustbowl and an Eden.

Reports started reaching us, one wet season, that the rains had failed entirely on the park's western boundary. These told of severe drought conditions in which the very young and very old had begun to die.

I was anxious to witness the effects of drought for, though undeniably distressing, it is a powerful force in the natural equation and must be reckoned with. So I promptly accepted State Veterinarian Ian Hofmeyr's invitation to join him on a fact-finding trip to the area. He wanted to collect and tabulate data on the condition of the larger herbivores and their habitat,

Right: *Growling ominously, two nomadic males dispute the remains of their kill. During the dry season, lions turn the prey species' dependence on permanent water-holes to their own advantage. Ambushes are laid and the hunting is good.*
Far right: *Stunned by the sun, webs of saliva trailing from her jaw, this lioness endures the midday heat. She pants at over 100 breaths a minute and loss of body fluid through evaporation keeps her cooler. By night her respiration will diminish to less than one breath every six seconds.*

to make recommendations for future policy planning. We packed supplies for a five-day camp and left immediately.

I was pleased to be making the trip with Ian. He typifies those men often found in remote places – intense, capable, comfortable with silence and at ease in wild country. He is withdrawn among strangers and seems unable to articulate his convictions without offending somebody. We understood each other and got on well. I admired his dedication, which he was at pains to mask behind proper scientific dispassion. I admired his driving less. He drove fast and I held on for dear life and only when we nearly left the road on one sharp bend did he slow down.

It was a long way to the affected area, much of it through stunted mopane scrub. At this time of the year, as the days grow longer and warmer and summer approaches, a new sound pervades the bush country – the high-pitched song of millions of cicadas. All day, and on into the evening, their cacophony rings out. They consort in colonies, so that as we passed, the shrilling came in peaks and troughs of sound. Responding to the mating instinct and to avoid predation, they slough off their old identity and in new, cryptically-coloured guise, skilfully meld against the gnarled bark of the mopane. I have found the abandoned husks of the original insect clinging to the tree trunks like diaphanous shells – frozen in attitudes of arrested motion.

Soon after sunset we reached the site we had chosen as a base. Since we both travel light, camp was established without delay and we settled round a low fire to grill a meal and relax. Insulated from Nature's extremes as we are, no time of the year is altogether bad. There are pleasures to be had even when the going is toughest.

Evenings around the campfire are a time of renewal, a time to sit and savour the light, refreshing breezes and the sounds and silences.

We had not long settled before monarch moths, with large false eyes prominent on their wings, ghosted down from the surrounding trees, attracted by the chilled white wine we sat drinking. They gathered at the

Top right: *The sun is the source of all life, but also fiercely sucks moisture from the grass, leaving it parched and bleached. To survive the desiccating winter months, the grass lies dormant, seemingly dead, but with the first rains it sprouts afresh.*
Right: *Through a shimmering heat haze, springbok listlessly nibble the dry grass. In winter the perennial grasses go to seed and, although much less nutritious than green grass, become standing hay to sustain grazers at a time when malnutrition is an ever-present threat.*

lip of the glass to probe with a delicate, curling proboscis. It was all they could ask for – sweet and cold and wet, although they had not reckoned on the alcoholic side effect. After drinking their fill they reeled off into the night, gloriously pie-eyed and not accountable for their conduct.

Early next morning we contacted the resident conservator who reported that the desperate situation continued to deteriorate. No rain had fallen and the game which had gathered at the man-made boreholes had devoured all available food.

Concentrations of animals at the boreholes were a normal development, but without rain there had been no plant regeneration. With no alternative surface waterpoints, the herds were unable to leave the vicinity of the boreholes, but were forced to range further afield each day in search of food. In their weakened condition they could hardly muster the necessary energy, and unless rain came soon, large numbers of antelope faced malnutrition and starvation.

We could do nothing except monitor the situation and hope for rain. The best way to assess the damage was to walk that parched, suffering country. Shouldering light backpacks, and in spite of a sun that roared overhead and the thickening heat as the day advanced, we set off at once. To the east translucent, pale grey cirrus creased the seared blue sky.

The devastation to vegetation was immediately apparent. Where grass still persevered all that remained was stubble – brittle and useless. Around waterpoints and other trampled areas of high game concentrations, only red earth and heat-reflecting rocks remained. Trees and shrubs were stripped, from below by kudu and eland, from above by giraffe.

Africa's wild animals have evolved over a span of 70 million years; during this long course adapting to the heat and rain, to poor soils and coarse vegetation. Drought was always an element to contend with, but the coming of modern man compounded the problem.

Left: *An alert, slender mongoose carefully surveys his approach to the water-hole. Should his presence be discovered by the large flocks of birds at the water, they will mob this inveterate egg-thief and devourer of their fledglings.*
Below: *The water-hole is both the centre of life and the scene of possible death. Great herds come to ease their thirst and predators come for the chance of an ambush kill.*

Etosha is no different from other wildlife reserves in Africa whose boundaries are artificial rather than natural. Such boundaries inevitably result in the disruption of many migratory patterns and the compromise of ecological balances. Age-old seasonal population shifts are abruptly halted. If wet season pastures fall within a game park, and fences prevent the herds from departing in the dry season, alternative water sources must be provided or they will die of thirst. However, when subterranean water is made available, the crowding this causes brings unnaturally high feeding pressures to bear on an environment never intended to support such numbers the year round. If not checked, this debilitating process soon creates a wasteland.

Before the arrival of technological man, wild animals abandoned a degraded area for better pastures, so allowing the land to recover. Today game-proof fences effectively prevent any such movement. Unfortunately the fences are necessary to deter wild animals from wandering on to adjoining farms where they are likely to be hunted or may transmit disease to domestic stock. The fences also curtail the activities of modern, motorised poachers. Because solutions will have to be found within the parks' perimeters, recognising and analysing ecological upheavals is vital.

To register the effects of the drought it was important to locate the carcasses of as many victims as possible. Here, death's beneficiaries, the vultures and marabous, were of great help. Circling in tightening spirals of descent, they signalled the whereabouts of each new fatality. There were far more carcasses than the vultures could dispose of, yet these birds, heavily gorged, shuffling and quarrelsome in spite of the largesse, were everywhere.

Early arrivals hissed and hopped with necks extended and wings outstretched, as menacing and hostile as a pack of street toughs. Others probed with old women's necks into cavities in the corpse. At our approach they moved away; heavy-winged, raising clouds of dust, they flopped into the air, trailing white pennants of excrement. In the trembling heat of the silence that followed, all that remained was the sweet stench of putrefaction – hanging close and thick and unmistakable in the air.

The appearances of the dead were striking; an ostrich, so long gone that the bones of its skeleton moulded themselves to the dry earth, odourless and innocuous; then the putrid remains of a kudu, the surrounding earth stained with exuded juices, flesh dripping, and the whole heaving with the movement of maggots; or an eland, newly dead, the ground scarred by the drag marks of its final agony, its mouth agape in mute appeal as excited flies gathered.

Far left: *In spite of the vastness of the Pan, this startled wildebeest holds to a familiar game-trail as it thunders off. Animals such as wildebeest are psychologically conditioned to cling to familiar routes and places, sometimes even to their disadvantage.*
Left: *A land monitor – an unblinking, imperturbable latter-day dragon – sprawls listless in the sun.*

I felt awed and reduced by death's profligacy. The wounded, impoverished land reeled under its dictatorship.

The eland in particular suffered terribly. Many of the cows, lacking milk, abandoned their calves. One mid-morning Ian and I came upon a herd drifting listlessly away from a borehole. The eland were in poor condition and walked with a purposelessness suggesting they had nowhere to go. A lone calf remained, sitting forlorn and without hope in the meagre shade of an acacia thorn. It bore fresh wounds from the kicks and butts of adults that had rebuffed its attempts to suckle. Yet the herd was all it had. It rose unsteadily to its feet and tottered a few following steps, but the effort required was too great. It swayed then collapsed, its legs gathered in under it, resigned, but seemingly at peace. It bawled once after the departing eland, was ignored, then fell silent.

We approached it, slowly and quietly to avoid adding to its distress, but by now all its remaining passions were directed inwards and it regarded us calmly, from bright, composed eyes, in which the last of its life forces had seemingly gathered. We had no choice, and as the sound of the shot died away, stared at the crumpled form at our feet and felt, in our helplessness, utterly downcast.

We stayed in the area for a week, and each day the pattern was repeated. The fierce sunlight drained everything it touched, burning all colours to an ashen glare. All day long it did not rain and at night it still held off. The harsh, sobbing air dried the membranes of our throats and noses; the death toll climbed.

At last we had to leave and on the road back we saw how close the rains had come. The humidity rose dramatically and though it was still hot and muggy where we were, the dark blue sky to the east looked cool and moist. Flocks of kites appeared, travelling ahead of the storm front, intent on harvesting the swarms of termites flushed by the advent of rain. A pair of jackals faced into the wind, eyes slitted, noses held high, sniffing.

Top left: *A white-backed vulture casts a baleful eye over this well-picked zebra rib cage. Vultures are among the few creatures that benefit from the agonies of drought.*
Bottom left: *A drought victim, whose only epitaph is the surrounding bare earth stippled by the foot-prints of feeding vultures.*
Below: *The drought progressed and hyenas, like ogres in a charnel house, relentlessly scoured the wounded land in search of fresh carcasses.*
Overleaf: *As we drove from the drought-stricken area, the long-awaited rain clouds were building up against the horizon, promising rain and the end of another dry season.*

The Wet Season

The whole tempo of life in Etosha is governed by the rains. Yet I would have been mistaken had I imagined that their arrival would be dramatic, matching the intensity with which they are awaited. The promise of rain is plain to see – flocks of heavy, black-bottomed cumulonimbi, lightning playing purposefully in their massy depths, gather along the eastern horizon. But they approach tentatively, reluctantly, as if uncertain of their welcome. During the day they are easily turned aside by unfavourable winds and the fierce summer sun. When the first raindrops eventually fall, it is at evening and into the night, probably because it takes the heat of the day to build up the clouds to a suitable state of instability. It is a thin, sour rain thrown down by fitful gusts of wind, but it nevertheless signals the turning of the seasons.

In the days that follow, the main body of cloud reinforces the vanguard, until at last a strong north-easter looses wind devils that raise towering dust storms across the pan and drives the swollen clouds forward. The humidity rises, thunder crackles, and forks of heat lightning flare against the bruised velvet sky. There is a tension and electricity in the air that communicates itself and sparks off responses. Stiff-legged, excited springbok stot with arched backs, prominently displaying tufts of erected white hair along the base of their spines. Dignified gemsbok forget themselves and cavort like calves, whilst overhead commandos of swifts rake the air in pursuit of swarming insects. Then the storm breaks with true African vigour. The rain comes hard with an odour of dust that has waited too long. In minutes the dried-up surface of the earth is no more – a sea has taken the place of the desert.

The wet season lasts from December to April, but the average annual rainfall is not heavy, ranging from 400mm in the west to 500mm in the east. The bulk of the rain arrives as torrential downpours, mostly lost as run-off which the newly-waterlogged earth cannot absorb. The amount of rain ultimately available for plant regeneration depends on the inevitable reappearance of the sun after a cloudburst. For the sun, which provides energy for growth, also takes its toll by evaporating nearly 80 per cent of the rainfall. It dehydrates soil and vegetation, yet, of necessity plant life has adapted to this harrowing climatic regime of desiccation and saturation.

Until now the land has waited, dormant but receptive. Where rain falls, smaller organisms respond immediately. Insects hatch, grasses germinate and thousands of moths take to the air. Awakened from aestivation, tortoises and bullfrogs reappear; and the land begins to change colour.

A green lawn of new grass stretches across the plains to the Ondundozonananandana hills that stand out sharp and hard in the clear light. Pink and white amaryllis lilies, and occasionally the superb gloriosa lily, add splashes of unfamiliar colour. Bright yellow tribulus weeds transform patches of bare earth in floral displays that are startling in this normally sombre-hued landscape. Orange mormordica creepers seek new frontiers, sending forth exploratory tendrils, and mopane trees replace their old russet foliage with new leaves of pink and lime.

I enjoyed being out on the plains in the rain among the herds as they hunched, faced away from the downpour. Sounds lapped in like gentle waves, suppressed and muted by the falling rain. After the storm, when the sun had reappeared, the light on the plains was as clear and true as a Vermeer. The damp earth steamed with a rich barnyard odour. It aroused subconscious memories of a childhood spent closer to the land than the years that came after – and, watching a lion shake the rain from his mane, I was pleased to be there. Nothing could have suited me better at that hour, and at that place and time in my life.

With the rains under way, their water collects in natural drainage lines, creating an instant habitat for waterfowl. Pans become lakes, unsightly gravel pits soften and metamorphose into spreading sweet-water pools. Now come hosts of ducks and geese, honking and fussing in restless excitement. Red-billed teal are most numerous. When startled they take to flight, sweeping by in a high-speed ballet, pinions whirring. Cape teal, maccoa and knob-billed ducks. Cape shovellers dip and up-end in search of food, and Egyptian geese are everywhere.

Sedge and reed thickets attract nesting colonies of yellow bishop birds. Darting males in jet-and-gold breeding plumage tour the dense stands of flooded shoreline grasses, selecting suitable sites. Nervous gatherings of terns collect on the beaches until alarmed, when they boil up like dark speckled leaves of natural cream or ivory. The insistent, mournful little whoops of a lovelorn hoopoe and the ringing, lonely cry of a single fish eagle.

Of the rain falling to the north and east of the park, much feeds two watercourses, the Ekuma River and the Omurumba Ovambo, which flow

Previous page: *With the rains, water collects in natural drainage lines, pans become lakes, and unsightly gravel pits metamorphose into spreading sweet-water pools. The only reminder that a few weeks previously this was a dust bowl is the large suspended communal nest of social weavers – birds of the dry country.*

Far left: *The female cheetah and her four cubs, caught in a thunderstorm on the Namutoni plains, look as miserable as only wet cats can.*

Left: *Sculpting soft runnels of mud, the night's torrential rains drain away before the dry African earth can soak them in.*

Top right: *Drowned termites float in a puddle left by the onset of the rains that triggered their nuptial flight. After heavy showers, sexually mature males and females pierce the surfaces of the great termitories that scatter Etosha's plains, and, almost in unison, millions grace the air in a brief flight that few survive.*
Bottom right: *Vigorous heads of Brunsvigia lilies splash unfamiliar colour on a landscape that had been parched only a fortnight before. After a brief, flamboyant display the flowers wither, to disappear as suddenly as they came.*

into the Etosha Pan. The waters of the Omurumba Ovambo first enter Fischer's Pan, a subsidiary of Etosha, where large flocks of flamingos, attracted by the brackish waters' rich algal broth, soon arrive. In high, jagged columns, muttering querulous exchanges, they circle the pan before deciding it is safe to land. Gaudy pink lesser flamingos strut and show off. Together with the paler greater flamingos they wade in extended lines across the pan, heads underwater, sweeping up-ended bills back and forth as they feed. Disturbed, they splinter into feathered fragments, displaying their roseate undercoverts, then group in flying formations to seek greater solitude.

Etosha lies on a major migration route, or 'fly-way', and large numbers of waders as well as other migrants from the north pass through at the beginning and end of summer. Most remain for only a few days or weeks before departing. The availability of surface water causes great fluctuations in water-fowl numbers. In a drought most species leave the area, although some may concentrate at the larger springs.

I spent one summer at Etosha when the rains were exceptionally heavy. Storms raged, the raindrops ricocheting off the packed earth like bullets, and the wind whipping the trees into a frenzy. So much rain fell that the salt pan became a lake again, revealing itself as it might have looked millions of years ago before evaporation dried it out.

On cool, grey mornings, the fresh wind – promise of more rain to come – raised small waves to lap the high ground of newly-created islands. It all seemed so strange, almost surreal, accustomed as I was to the arid, dust-bowl conditions of winter. From one of the islands three ostriches, birds I have always associated with hot, dry places, unhesitatingly waded chest deep, tails arched to prevent soaking, across a channel to reach the mainland. A herd of gemsbok crossed to the island, perhaps recognising in its isolation a refuge from predators.

The warm, nutrient-rich waters teemed with fish, particularly barbel and tiny silver tilapia. Herons, egrets, marabou and yellow-billed storks gathered to harvest the windfall, converting fish into energy then reintroducing it into the environment as guano at a prodigious rate. Although not equipped with a bill designed to process large fish, a sacred ibis joined the throng and did remarkably well. Aggressive grey-headed gulls attempted to pirate the catches of any easily-intimidated fisher. As the waters of the pan evaporated, a heavy smell of brine hung in the air, an estuarine smell like freshly-caught crayfish sniffed at in a harbour market-place. Stepping carefully, marabous picked at flapping barbel trapped in the turgid, brackish water that remained.

Summer is the time for extravagant displays on the part of the many sexually-aroused male birds. Splendid cock paradise whydahs outrageously flirt their trailing black tail feathers. Drab-coloured, red-crested korhaans concern themselves more with acts of dare-devilry. Soaring vertically upwards, the male stalls at his apex, tucks in his wings and plummets to earth in a daring free-fall, only to level out at the last moment and

Top: *This strutting male kori bustard's courtship display unequivocably announces his amorous intentions far and wide. Tail raised to reveal white undercoverts that flash in the sun, he inflates his throat sac until, just before it seems ready to burst, he opens his beak then snaps it shut, producing an extraordinary booming gulp.*
Centre left: *The tail feathers of a red-billed teal protrude from the mouth of this enormous rock python; the rest of it bulges his neck. Pythons are as much at home in water as on dry land, and this particular snake fed almost exclusively off the flocks of teal that frequent Klein Namutoni Spring.*
Centre right: *Perfectly reflected in the still waters of Klein Namutoni Spring, a stilt delicately probes the shallows for insect larvae and small crustaceans.*
Bottom: *Over-extending itself in the face of such plenty, a very determined sacred ibis considers how best to manoeuvre this large barbel into its mouth. Its bill is not designed for prey this size, but after several attempts – egged on by piratical grey-headed gulls – the ibis eventually swallowed it.*
Top right: *One of the features of the wet season is the extravagant displays of courting birds. Here blue cranes joyfully celebrate their union with an elaborate dance, orchestrated by their loud, distinctive, bugling calls.*
Far centre right: *The moment of lift-off, as this white pelican flaps from its temporary sanctuary on one of Etosha's reed-shrouded artesian springs.*
Bottom right: *Its iridescent wings and patent-leather bill glowing in the light of late afternoon, an African shelduck bursts into flight.*

settle nonchalantly. What timorous hen could fail to be impressed by such a rugged fellow? The arrogant kori bustard struts back and forth, tail raised to reveal his white undertail coverts. The glistening white spot flashes like a heliograph in the sun, and by inflating his throat sac he produces a reverberating boom that unequivocally broadcasts his intentions.

Even well into the rainy season, with many millimetres on record, sudden fresh showers are never taken for granted. Each seems unique and is always appreciated. If I happened to be walking through close mopane forest when new rain came, it would announce its arrival with a small song – a thin breeze rustling the high branches. Then the temperature would drop, the wind rise and the first raindrops, plump and self-important, would fall.

The forest is a strangely quiet place, not given to loud exclamations, and if the only sound was the cooing of a turtle-dove, with the onset of rain it too would fall silent, as if considering. Then it would call again, the same tune but with an added note of celebration.

The rains determine the seasonal migrations of the plains game. With the wet season under way, herds move from their dry-season pastures, following the grasslands edging the southern periphery of the pan, and on to the acacia-studded savannas north-west of Okaukuejo. It is a determined exodus, with family groups of zebra and herds of wildebeest congregating in large numbers to make the trek, and tail-wagging companies of springbok following close behind.

In none of the larger game parks in southern Africa is one made so aware of sudden shifts of great numbers of animals as in Etosha. The impact of the trek is dramatized by the vast open spaces of the plains and the fact that much of the travelling is done by night. One day the plains ripple with feeding animals; then, within hours of the first thunderstorm, are left achingly empty. There is an urgency to the herds' movements; they have more than 150 kilometres to travel to their prolific summer pastures.

The pull to the north-west is not stimulated so much by a need for rain as drinking water, for it also falls on the winter grazing areas at this time. The movement is away from the tall perennial grasslands to short annual grasslands where the rain has promoted new growth. Zebra and wildebeest prefer certain grasses over others, but basically what they appear to seek is short, growing grass which is easily digested and most nourishing. This use of short grasses at their most nutritious, is a form of natural rotational grazing practised particularly by zebras.

Top: *An inquisitive black-faced impala fawn peers from the summer lushness of its woodland habitat.*
Left: *I enjoyed being out on the plains in the rain, amongst the herds as they hunched, faced away from the downpour. The sounds of the place lapped in like gentle waves, suppressed and muted as they were by the falling rain.*

Top left: *A column of greater flamingoes, honking querulously, circle a pan to satisfy themselves it is safe to land. In years of exceptionally heavy rainfall, Etosha becomes a breeding-ground for more than a million of these summer migrants.*
Left: *Tracking a storm, this restless herd of gemsbok move to the nutritious flush of new growing grass which local showers have produced.*
Above: *Lapped by the floodwaters of Etosha Pan, an island creates an isolated refuge for this clutch of avocet eggs. Though the eggs are laid in the open, their cryptic colouring increases their chances of survival, for their ground colour matches their surroundings and a myriad of flecks and spots disrupts the tell-tale egg shape.*

Impressive as modern-day migrations are, they represent only a fraction of the herds' movements before fencing, hunting and ranching contained them within the park. The animals' use of smaller grazing areas for longer periods has allowed parasites and bacterial diseases to prosper, their pathogens building up at such sites as permanent or seasonal waters, where disease can reach epidemic proportions among stationary game populations.

Anthrax epidemics coincide with the rainy season. With rinderpest, anthrax is endemic throughout much of Africa, and can become rampant, its bacteria affecting warm-blooded species and resulting in a fatal septicaemia.

In Etosha the preconditions that encourage anthrax have increased dramatically over the past decade or so, with development schemes within the park unwitting contributors. A decision to improve the park's network of roads with all-weather gravel surfaces resulted in excavations which filled with rain-water during the wet season. Their lower salinity led to these gravel pits becoming favoured drinking places and, because they were impervious and deeper, the pits retained water up to five weeks

Left: *Reaching for the sun, a* Moringa ovalifolia *seedling thrusts from the fissure of a dolomite boulder. If it survives and matures, its stem will mould and accommodate itself to its cramped environment.*
Bottom left: *A knot of bachelor kudus relax in a field of buttercup-yellow* Tribulus zeiheri. *Until they seed and die, these annual plants both decorate the previously bare landscape and provide useful fodder.*
Below: *Like some prehistoric monster at the end of its era, this aroused, puffed-up, hissing, tongue-probing land monitor is not poised to attack – instead it is doing all it can to avoid a fight. The aggressive display is intended to warn off any would-be antagonist.*

Below: *Overhung by a large communal spiders' nest, a black-backed jackal drinks from a fresh rainwater pan. These life-giving pans also act as anthrax incubators and, although the jackal is immune to this disease, he helps transmit it by feeding on infected carcasses and then drinking.*
Right: *Sweeping in from the sky, a white-backed vulture drops on an anthrax-infested elephant carcass. If a diseased elephant dies near a water-hole, other elephants desert the area and avoid drinking from that particular source for several weeks.*
Overleaf: *Dwarfed by towering thunderheads, a herd of giraffe move onto the Halali plains in the early evening.*

longer than the rain-water pans. This encouraged the migratory herds to remain for longer than previously, creating extensive grazing pressure, over-trampling and range regression. The resulting contaminated soil and water provided fertile breeding grounds for anthrax – stimulated by high summer temperatures. And with wildlife remaining longer at these sites, susceptible animals were more prone to infection.

Etosha's first recorded serious anthrax outbreak occurred in the Namutoni area during an exceptionally good rainy season, when at least 240 animal carcasses were incinerated on the spot to prevent further spread of the infection. Its source was traced to a large gravel pit and, after this had been disinfected and filled, only sporadic cases occurred. So, though anthrax can probably never be eliminated from the park, the most effective long-term control seems to be to dose artificial water-holes in the overcrowded and over-used areas.

Anthrax is the sting in the tail of the rainy season. The summer months create the favourable conditions needed for mothers to produce their young, but also release the seeds of death that leave some of the newborn as orphans to starve or fall prey to predators.

This may appear a capricious and cruel system, but the coming of the rains provides enough compensation. Life is renewed, but a price is exacted. The fate of the individual is of no consequence as long as the status of the species is maintained. If infectious diseases are contained the nett result is the maintenance of a harsh ecological balance, which is all that Nature requires or expects.

A Home on the Plains

The land is at its nutritious best during the rainy season, setting the stage for the arrival of thousands of newborn animals. Their birth at this time of year provides young herbivores with abundant green forage at a critical stage in their development. By the time they start eating solids, there is a ready supply of tender, growing shoots of grass.

Most of the ungulates give birth during a well-defined period which coincides closely with the rains. When rainfall is late or insufficient to encourage new grass to grow, pregnant females can, to an extent, delay labour until conditions improve. Birth peaks may also differ geographically within the park. In eastern Etosha where drenching rains usually fall earliest, wildebeest and springbok begin to have young as much as a month earlier than those around Okaukuejo. Seasonal breeding is generally more pronounced where climatic changes are extreme, as in Etosha, than in temperate regions.

Discrete birth seasons, with most young being born over a four-week period, are particularly characteristic of springbok and wildebeest and, to a lesser extent, zebra. This intensive synchronization of births seems designed to ensure the highest survival rate for species whose young are particularly vulnerable, either because they are conspicuous or because the social system does not adequately provide for their protection. Predators are so inundated by potential prey that they cannot harvest it all. This glut for a limited period results in a lower mortality than if births were spread throughout the year. There is an almost 100 per cent mortality for those born before the peak, when a single newborn is highly visible.

With so many females having young at the same time, and usually during the daylight hours, it might be assumed that births would be easy to observe. But this is not the case. The act of giving birth is of very short duration and, being so vulnerable, the mothers-to-be are naturally wary and difficult to approach.

I was fortunate in witnessing either part of, or the complete births of a number of springbok, zebra and wildebeest, though not always being able to get close enough to take photographs. The mother can forestall parturition for several hours if the circumstances are not propitious.

In the case of springbok, I learnt to detect a female's characteristic behaviour immediately before birth. As an ewe feels her time approaching she separates from the rest of her herd, so that my first clue was an obviously pregnant solitary female. At this initial stage the only other distinguishing feature was a slight elevating of her tail.

During the first complete birth I observed, this behaviour pattern was maintained with very little variation for an hour or so before more obvious symptoms of approaching maternity were signalled. These included holding her tail stiffly erect, nudging and licking her hindquarters and, periodically, stretching with forelegs and hindlegs widely spread. During this time she roamed extensively across the plains, appearing to graze without doing so; sitting only to restlessly rise almost immediately. Although I circumspectly kept my distance and did not arouse any apparent distress in the ewe, I could never be sure what effect, if any, my presence had on her. Of all the births I witnessed, I am inclined to believe that I interfered in no way, as mothers can discontinue the birth process up until the time the head first appears, and in no instances did they attempt to do so.

Eventually, she settled in the meagre shade of a littoralis aloe, and almost immediately contractions started, her neck tendons bulging with the effort. Soon afterwards the amniotic fluid was ejected and five minutes later the taut white balloon of the foetal sac emerged under her tail. With nervous, undirected energy she shifted position, getting up and walking about ten metres before settling in the dusty depression of a zebra wallow. During a prolonged wave of contractions which followed she sometimes gave little gasps of exertion. After 25 minutes the fawn's forefeet appeared. She rested for about ten minutes and was again wrenched by a series of spasms. The fawn's head appeared on the wave of an enormous contraction. She relaxed and contented herself with occasional sniffs at the protruding embryo, still enclosed by the foetal sac and registering no signs of life.

Once she stood and seemed about to move off but, after distractedly fidgeting for a moment or so, hesitated, then lay on her side, with outstretched legs and her head on the ground. The shock of subsequent contractions arched her body, her head and legs now rigid in the air. The shoulders of the fawn were pushed free and it shook its head groggily in its first sign of life. The ewe attempted to lick it, but could not reach. Again she reclined and in two heaves got the body and hips out. In the final stages the fawn delivered itself, wriggling and pulling its hind legs clear. Without getting up the exhausted mother reached around to lick and nibble at the remains of the sac.

Finally she rested, watching the fawn as it made its first shaky attempts to stand. Within a few minutes the mother stood up and moved to her baby, nudging and licking it encouragingly. Inherent wisdom and its mother's insistence forced the fawn to attempt to stand. Bracing its hind legs, it managed to raise its rump – and promptly collapsed. It continued trying until, twitching and trembling with the effort, it stood with stick-like legs splayed until it toppled in a tangle of limbs, to relax and gather its reserves for one more try. This time is succeeded. Unable to lift a foot, it stood swaying, its head lowered. Then it cautiously started feeling its way around its mother and, when it found her teats, started suckling.

At the same time a ram and a yearling approached them from the nearby herd. When the fawn had finished drinking it lay flat, completely hidden among dry salt bushes. The ram and yearling sniffed at the mother, the ram showing the Flehman grimace associated with sexual arousal. The yearling inspected the fawn while the mother, tail stiffly extended, moved a little way off, followed by the ram. After walking about 150 metres from the fawn, the male, who had continued to present the proud posture and curled upper lip typical of the breeding stance, attempted to mount the ewe. She discouraged him by continually moving forward and, when he had accepted her rejection, started feeding. Twenty minutes later she returned

Previous page: *At my approach, this nursery group of young springbok fawns nervously flattened their long, conspicuous ears, hoping to avoid detection.*

The birth of a springbok is more difficult to photograph than one would believe though many hundreds may be born in a period of about two weeks. The ewes are wary and the actual delivery is very quick. Minutes after this ewe started her labour, the front feet of the fawn emerged (Top right) *soon to be followed by the head and shoulders. A pause, another contraction and the birth was complete. Strength flows into the fawn as its mother nibbles at the foetal sac* (Centre right). *Despite her obvious exhaustion, the ewe was anxious to encourage her offspring to its feet, nudging and licking it to greater effort. The fawn lies sprawled awkwardly in the open, but it has no scent and, as long as it remains still, is perfectly safe* (Bottom right). *Unsteady, but on its feet at last, the still-damp fawn surveys its world for the first time* (Far right). *The long, white, silky hair at the base of its spine is particularly conspicuous and will act as a visual aid between herd members on the run.*

Left: *With its foetal membrane still adhering, this wildebeest calf is ready to stand. In the first moments of life all its energies are directed towards following its mother back to the sanctuary of the herd. Learning to walk takes less than 10 minutes.*
Below: *This wildebeest cow, closely followed by her new-born calf, left the scene of the birth as soon as possible. She will only expel the dangling placenta some time later, giving the calf time to develop its strength before the bloody afterbirth attracts the attention of scavengers and predators.*

to the fawn, the ram still following. The fawn stood up and suckled once more while the ram sniffed it casually before rejoining the herd. The new mother and her fawn followed minutes later, the fawn no longer showing any signs of unsteadiness.

Isolation during birth is extremely important to both the ewe and her offspring as there is a mutual imprinting of sound and smell at this stage. Fawns, as soon as they are completely steady on their feet, spend much of the time with others of the same age in large nursery groups. If danger threatens, the females rush to the nursery, each mother assuming responsibility for her own offspring's safety.

Fawns remain in their mother's custody until the following year when the females give birth again. The yearlings will continue to remain in the herd, often associating closely with their mothers but gradually assuming greater independence.

One of the most rewarding aspects of spending long periods in an undisturbed natural environment was the opportunity it afforded to acquaint myself with the habits and customs of the wild animals, part of a rich natural tapestry, intricate and complex – and just as dramatic as hunters' tales or Hollywood myths.

Some species are shy and reluctant to reveal themselves, and it was only with patience, some knowledge of their behaviour and a measure of good luck, that I got to know them. Particularly pleasing and satisfying was the opportunity I had to follow the progress of two parent bat-eared foxes as they raised a litter of four pups far out on the Charitsaub plains.

I discovered the family towards the end of what had been a very hot November. Early one morning I noticed a flurry of dust at the centre of which a bat-eared fox was digging furiously to excavate and enlarge the burrows of an abandoned ground squirrel warren. I knew that jackals, Cape foxes and bat-eared foxes all pupped at this time of the year and guessed that it might be remodelling the ground squirrels' old home as a nursery. My guess was confirmed when he stopped digging and backed up, allowing a female to scramble out of the burrow. I could see from her enlarged teats and the surrounding damp fur that she had recently nursed pups.

The two adult foxes were pretty little animals. Their soft, thick fur ranged in colour from brownish grey on the back to buff on the flanks, throat and belly. Their delicate, pointed muzzles were light brown, darkening towards the eyes. Their foreheads were black, as were their chins and feet. The most conspicuous feature, however, was their enormous ears, more than one third as long as the animal is tall. The fluffy coat, winsome face, and preposterous ears gave each a delightfully cuddly appeal.

Top left: *Every year thousands of young mammals are born in Etosha. Some reach maturity, some do not. Among the latter was a steenbok fawn caught only days after it was born and now belligerently guarded by a martial eagle.*

Bottom left: *This young kudu has successfully survived all the dangers of growing up in the wilds and will soon leave its mother's side as an independent adult. Though still suckling, this is probably more for a sense of security than for sustenance. Of its first months much was spent lying concealed well away from its mother, and suckling helps maintain the mother-calf bond.*

Below: *Not all births are happy occasions. Due to a malposition, this springbok lamb died before birth and the ewe is wracked with muscular spasms as she attempts to deliver herself of the foetus. If she is unable to do so before nightfall, she will present an easy target for any predator.*

Right: *I was fortunate in witnessing the birth of this zebra, for in Etosha most foals are born in the dark of pre-dawn. The mare was totally unconcerned by my presence as she encouraged her youngster to stand.*
Far right: *The foal on spindly, delicate legs splayed for support. Though soon it will be introduced to the rest of the herd, at first its mother keeps other members at bay lest they interfere with the imprinting of her image on her foal. Without this vital bonding they may not be able to find one another in the milling herd.*

However endearing their appearance, it is the result of practical adaptation to the rigorous environment. Bat-eared foxes favour dry open country and even prosper in deserts. They can withstand high daytime heat and tolerate rapid and extreme temperature changes between day and night. The prominent extremities, such as the pointed nose and huge ears, effectively radiate excess body heat, while the long guard hairs of their fur cover a dense, woolly undercoat which serves as insulation against extreme temperature variations. Their water requirements are very low because of physiological adaptations to their arid environment, and they are able to extract sufficient moisture from their diet of insects, rodents, the nestlings and eggs of ground-nesting birds, wild fruits, tuberous roots and other vegetable matter.

The female seemed larger than the male and, being better groomed, was the more attractive. The male's ragged fur was clotted with burrs and grass seeds. This matted coat may have made him look smaller than his fluffed-out companion, and his generally disreputable air was not improved by the two canine teeth that protruded incongruously from his upper jaw. But he was a solicitous and loving mate, cosying up against the female when she flopped into the shade of a stunted shrub, energetically grooming her with his incisors.

Although at first the pups did not appear, I was pleased that the parents took little notice of me. Ordinarily bat-eared foxes have no use for inquisitive humans, but these two had obviously become accustomed to tourist vehicles, and were later to communicate their lack of concern to their pups. Because of their indifference I was able to park within 25 metres of the burrow. It was important not to disturb them as, apart from any other considerations, they were liable to move to another burrow under cover of darkness, and I would probably lose them.

Eventually the female moved off. The male crossed to the burrow and, peering down, gave a short thin call. Almost instantly four tiny pups rushed out, tumbling over themselves in their excitement. On seeing me

Top left: *The bat-eared fox pups seemed to regard this grooming session by their father as a mixed blessing – enjoying the attention but rebelling at the thoroughness. The months we were able to spend observing this family of bat-eared foxes was one of the highlights of our many experiences in the wilds.*

Bottom left: *Still too young to accompany their parents on nightly foraging expeditions, the pups collect at the burrow entrance to watch the adults fading from sight into the gathering gloom.*
Above: *The madcap behaviour of the pups was reminiscent of kittens – rushing about with bunched backs, arched tails, scampering over the home site, threatening, then retreating, packing an individual, then indiscriminately switching allegiance.*

they stopped in their tracks and seemed about to retreat into the safety of the burrow when the male called again, and, trusting his judgement, they hurried to him. First they pushed at his cheeks and muzzle with their raised faces, then rushed underneath and butted his belly, wanting to suckle. He discouraged this by sitting, then set about grooming each in turn. Knocking them off their feet, and rolling them either on to their backs or sides, he nibbled vigorously, his ears laid back over his neck. The pups seemed to regard their grooming session as a mixed blessing, enjoying the attention, but rebelling against the thoroughness. Pinned by the weight of their father's nose, they first lay compliantly, then pawed the air and, emitting a high chittering call, attempted to escape. Usually they succeeded in breaking away, whereupon the male would pounce upon another candidate whether it had been groomed or not.

I estimated the pups to be about two weeks old when I first saw them. They were still wobbly on their feet and very unsure of themselves above ground. At the slightest untoward movement, sound or shadow, they would scramble frantically to the shelter of the burrow. They were miniatures of their parents though they had not yet developed the exaggerated ears and elongated muzzles. They were also much lighter in colour, a tawny buff, but already exhibited the characteristic dark dorsal stripe from nape to the tip of the tail.

Bat-eared foxes are nocturnal, another useful adaptation in conserving energy and fluids in a hot, dry environment. During the day they shelter from the sun in the deepest shade available and only move if disturbed. To photograph and observe them I left my house before dawn each morning to arrive at the burrow as the adults returned from a night of foraging.

My route took me through a mopane forest, the car's headlights illuminating startled genets and hares, then out on to the plains just as first light bathed the sky, firing the clouds close above the horizon with wild and splendid shades. Jackals hurried out of my way, stopping at a safe distance to stare over their shoulders, watching me pass. Herds of zebra and wildebeest, plodding their migratory paths, ignored me. Once, out of the corner of my eye, I saw a shifting of light from one shadow to the next and, turning quickly, saw a leopard glide across a clearing, as mysterious and insubstantial as a ghost.

The burrow was an unadorned shaft set on a stretch of plain dressed with a swathe of newly-erupted grass. Without the natural signposts I had noted along the route, I would have been hard put to find it again each morning. Occasionally when I arrived there was no activity at all in the neighbourhood of the burrow. Then, as the land warmed, the pups would show themselves at the entrance. After staring at me and reassuring themselves that all was well, they would emerge, stretching and yawning, to collect at the mouth of the burrow, sprawled on top of one another, enjoying the play of the sun. Then from across the cropped plain the mother could be seen approaching, and as she drew near the pups mobbed her joyously, beside themselves with excitement at the prospect of suckling.

Above: *This forlorn jackal pup howls its dismay at being left alone for the moment by its parents on the wide open plains.* Right: *The male bat-eared fox breaks off his grooming, his attention alerted by an unusual sound. However endearing his appearance, it is the result of practical adaptation to his rigourous environment – his huge ears act as radiators that disperse body heat, as well as being hypersensitive hearing aids.*

The female stood stoically during the assault on her teats, as the pups rose on their hind legs to drink, lunging and tumbling in their frenzy. Usually suckling continued for three minutes before the mother pulled away, discouraging any persistent pups with a growl. She would lope off to the shade of a shrub, distant enough to dissuade the pups from following and attempting to suckle again. Left on their own, the pups amused themselves in vigorous play. Their madcap behaviour was reminiscent of kittens – rushing about with bunched backs, arched tails, scampering over the home site, threatening then retreating, packing on an individual then indiscriminately switching alliance to turn on an erstwhile comrade. Their rough and tumble games were delightful to watch. All very young animals are appealing, but the bat-eared fox pups with their repertoire of antics, bright-eyed enthusiasm and irrepressible *joie de vivre* were in a class of their own.

Their lunatic rushes and seemingly unconcerned sparring, bewildering to their father as he tried to maintain order, were actually establishing a dominance hierarchy later to be reflected in their adult relationships.

As the pups developed and were weaned, their play began to imitate more clearly the foraging tactics of their parents. They would course the bare earth, their noses close to the ground, intermittently cocking their heads to one side and listening in a comical attitude of concentration. Occasionally they would start digging, but with little effect. The mother continued to keep her distance, and most of the pups' sustenance was provided by the male. He worked tirelessly into the heat of the late morning, unearthing tidbits to satisfy their demands, periodically dropping to rest in any available shade.

The male concentrated on unearthing as many grubs of the dung beetle as he could find. These armour-plated beetles are found in immense numbers wherever herds of grazing animals congregate. They deposit their eggs in the balls of dung, several of which they bury together. The bat-eared fox could hear the squeaking sound made by the larva moving in its hardened dung casement, dug up the ball, broke it open and secured the grub.

The male carried the intact grub over to the burrow, whereupon the pups would rush out and hungrily push at his jaws with their muzzles, encouraging him to release the morsel. As soon as he dropped the heavy yellow grub, it was snatched up by one of the pups who immediately retreated to the burrow to eat in solitary peace, leaving the harassed male to face the begging pleas of the others. Disconsolately he would wheel in his tracks and trot off, then repeat his search. Only once did I see him, with a grub held tantalizingly between his jaws, glance at the burrow and seeing no movement at all, surreptitiously swallow it. He then guiltily hurried to the next site, and having procured a grub, passed it on to the pups.

Bat-eared fox litters number from three to five pups who, after maturity, stay with their parents for several months, perhaps eventually making up part of an extended family. Throughout their lives they are fair game for all kinds of predators. I saw the male fox drive a black-backed jackal away from

Only a few weeks old, the bat-eared fox pups accompanied their parents on their first foraging expedition out on the plains, in the soft light of early morning.
Overleaf: *Shortly before being introduced to the rest of the herd, this two-hour-old springbok fawn is thoroughly cleaned by its mother, and her fawn from the previous year.*

the home site, but also saw, at Namutoni, a jackal trotting along with a dead pup between its jaws.

Larger birds of prey were a constant threat of which the pups were well aware. They would automatically scurry towards their burrow if the shadow of a circling crow swept the ground, and only later came to realize that crows would do them no harm. It has been suggested that the leopard, which undeniably has acquired a taste for other smaller carnivores, might be the greatest predator of bat-eared foxes. Whilst this may be true in other parts of Africa, the different habitats of the two species in Etosha seems to preclude a significant predator-prey relationship.

Wildlife photographers, Des and Jen Bartlett, saw a lone cheetah feeding on the Charitsaub plains. After it had abandoned its kill and a lappet-faced vulture had dropped to the spot, they investigated and found the remains of an adult bat-eared fox. The Bartletts also witnessed subadult cheetah cubs harrying and biting a fleeing fox which later proved the mother of a litter being raised in a communal den. The mother sustained mortal wounds as a result of her encounter with the cheetah cubs, and her pups, the youngest born in the community and now unprotected, were attacked and eventually killed by the neighbouring pups, apparently without intervention on the part of the remaining adults.

Yet it is probably not predation and its consequences, but disease, that most severely limits bat-eared fox populations. Disease affects their proliferation in a pattern typical of social, den-dwelling foxes. Bat-eared fox populations in Etosha noticeably increase until, reaching an undefined peak, they suddenly crash, possibly decimated by distemper. After such a disaster their numbers gradually rebuild until disease attacks again.

Without knowing it at the time, the last day I spent with the foxes ended on a cool, rainy, overcast afternoon about six weeks after I had first encountered them. The whole family had collected around the burrow, the gambolling pups challenging and feinting, their parents looking fatigued and uninvolved. A sudden loud peal of thunder frightened the pups sufficiently to send them scampering underground. They soon re-emerged looking prepared for anything; but by now gusts of wind swept the plains. The adults huddled together, crouched close to the ground, with their ears flattened. Then, as suddenly as it appeared, the storm lifted. Although no rain had fallen, the adults shook themselves as if anticipating the consequences of a drenching. The male distractedly groomed a propositioning pup, as the female trailed away into the gathering dusk. He watched her go, then quietly followed. The pups were two months old by now and anxious to accompany their elders further afield. The male seemed reluctant to have their company, but they were adamant.

My last sight of them was of the male trailing across the plains with the pups resolutely bringing up the rear. They disappeared at the point where the night sky meets the earth and did not return to their burrow. And though I would like to think they are content wherever they are, I never found them again.

Under a Hunter's Moon

In all the time spent in the African bush and after countless encounters, I have never learnt to take lions for granted. I doubt I ever will. Even when at ease a lion has an aura of power and leashed violence that implies a deep sense of menace. It is a predator epitomized, from the calm authority of its bearded head, in the determination of its pale gold eyes to the black tufted tassel of its tail.

The lion's mystique is emblazoned on the heraldry and psyche of our civilization. Few wild animals have made more of an impact on the traditions and myths of the peoples it encountered. Our ancestors respected and feared the lion, some even deified it.

Mankind's relationship with lions has undergone several changes since then, but whether reviled as a bloodthirsty man-eater, or idealized as companionable and kind, the fascination remains.

I remember my first Etosha lion with a clarity as vivid as if I were seeing it now. Soon after sunrise on my second day there, on a cool August morning, far out on the plains. A herd of zebra near by displaying a nice sense of community, indulging themselves in much nuzzling and rubbing. Larks talkative beyond us and a black korhaan shrieking at some personal matter. Then from out of the low sun, through parched winter grass, a single lion padded towards me. Where the diffused light touched his mane it took fire. His passage drew a dark wake through the silvery grass. He was staring beyond the car into the distance, and if he knew I was there he gave no sign. He paid no heed when the zebras snorted and bolted, striding purposefully he passed on by, heavy muscles flowing under his skin. Where he was heading with such deliberate single-mindedness I have no idea. He disappeared off the plains just as surely as he had come, without sparing me so much as a glance.

Since then I have come to know the lions of Etosha in many different situations and moods, in good times and bad. I have watched cubs grow from awkward playfulness to become consummate hunters, and have passed the time of day with the voluminously-maned males, so assured in their indolence. Some of my most pleasant memories are of evenings around a camp-fire, with lions filling the night with passionate roars.

Lions only fully reveal themselves at night, when they slough off the lethargy of day and determinedly attend to their nocturnal activities. As the sun slips below the horizon the cubs play more vigorously and lionesses stretch, yawn and rub cheeks, restless to start the hunt. Most hunting is done at night, although under favourable conditions they may also hunt by day. Unquestionably, however, lions stalk with a greater chance of success under the cover of night. They seem well aware of the advantage darkness affords, particularly out on the plains where there is little or no vegetation to screen them.

If the pride has spent the day resting in the vicinity of a spring, most, and usually all, will drink before starting out. The sound of their lapping carries clearly on the dry, still air. Before the last of the pride has finished, a lioness moves off, followed in an extended single file by the other females with the

cubs trailing behind them, occasionally running beside or around them, while the plodding males bring up the rear. With the moon staring in silence at the earth, it is moving to watch their shadowy forms merge and disappear into the anonymity of the flat, silver landscape.

The Etosha lions have acquired a reputation as enthusiastic man-eaters whenever an opportunity presents itself. They are probably no more deserving of the stigma than lions elsewhere, but the label persists. Their infamy is mainly attributable to an incident near the Okondeka contact spring in 1950.

A party of five itinerant Ovambos had camped beneath a large thorn tree near the water. At some stage during the night they were taken unawares and attacked by a number of lions. Only one of the men managed to escape into the safety of the branches where, bleeding and half demented with fear, he spent the rest of the night listening to the lions chewing on his erstwhile companions. He was rescued the following day by a passing police patrol and his gruesome tale has since been enshrined in Etosha lore.

Although since then no-one has been killed by a lion, lions tend to react aggressively to someone on foot or on horseback. While I was in Etosha different conservators had to fire warning shots to discourage what they considered determined charges. On the other hand, the noted ethologist George Schaller found that on the dozen or so occasions that he encountered lions while hiking in his Serengeti study area, they bolted at a distance of 30 metres or more. Once, however, a young male attacked his car, biting two holes in the fender. Clearly there are no hard and fast rules when dealing with lions.

By and large, however, lions in game parks open to tourists pay very little attention to motorized traffic. Provided a slow, cautious approach is made, it is possible to get within 25 metres or closer without disturbing them. If the lions I encountered showed any signs of nervousness I would immediately switch off the engine, giving them a chance to settle down again. They are generally reluctant to move when resting or feeding and, with a show of good manners, I have spent many hours in their company.

Indeed, some individuals became so used to my presence that on the several occasions I joined them at a kill on a treeless savanna they resorted to the shade cast by my car as the day warmed up. Once four lionesses crawled half-way under the car when the noonday sun stood directly overhead. I was glad to be of service, a small enough return for all the past pleasures they had given me.

Previous page: *Lions are the night-shift workers of Etosha. Certainly night gives them better cover, better chance of success . . . under a hunter's moon, their shadowy forms merge into the anonymity of a flat, silvered landscape.*

Far left: *Almost anything is fair game for a lion, but this frustrated lioness gnawed without effect on a leopard tortoise's carapace.*

Top: *A family of lions crowd flank to flank round the bloody ribcage of their wildebeest kill. The pride has no rigid feeding hierarchy so that each individual must join the shambles to maintain its rights and defend its share.*

Centre: *Ignoring the importunings of her cub, a lioness gnaws at the little meat remaining on the wildebeest skull she has claimed. Lions are most successful on a hunt when members of a pride co-operate, but at mealtime all co-operation ceases and it is each for itself.*

Left: *I came across this cub chewing contentedly on a bone, and watched as it cuffed and snarled at any adult that tried to take it away. When food is scarce, the cubs are the last to feed and the first to suffer. Many die of starvation.*

Right: *Lion cubs toy with the carcass of a young spotted hyena killed the night before by their mother, who had crushed its neck with one bite. Only recently has it been realised how many predators are killed by other predators.*
Below and below right: *Grimacing her distaste, a lioness looks up from mouthing a brown hyena she had paralysed with a vertebrae-damaging bite only moments before. The attack on the hyena appeared utterly gratuitous, for the lioness ate none of it and, after worrying it for a few minutes, wandered away.*

There are still lions at Okondeka and one grey, overcast morning I witnessed two of them take part in a remarkable example of interspecific predator aggression. As I drove up to the spring, the first thing I saw was the distinctive sloping back, shaggy coat and large pointed ears of a brown hyena. The brown hyena is a timid, nocturnal scavenger, rarely seen by day – the dull weather probably accounted for its activity at that time of the morning. Brown hyenas are more solitary than the spotted, even more bizarre in appearance, and lack their characteristic cry.

Having had few opportunities to observe them in the past, I watched with interest through binoculars as it investigated along the edge of the spring, for the most part ignored by the drinking herds. Finding nothing of interest it shuffled up an incline without noticing the two lions that lay at the crest. The lioness was aware of the hyena however, had kept it under casual observation until now and, as it drew nearer, she flattened into an alert crouch. The male lion looked up, interested in what had aroused the lioness, but immediately ducked his head on seeing how close the hyena had come.

The hyena held to its course, never for a moment realizing its danger; so preoccupied that it failed to see the lioness snake menacingly forward. It knew of nothing amiss until the lioness exploded from cover, then had only time for a hopeless, placatory cringe of submission before it disappeared under the impact of the charge. The lioness's first bite damaged its vertebra, partially paralysing it without killing it. The male hurried over and snatched up the limp, yielding body, worried it, then dropped it again. The lioness continued to mouth it for a few minutes longer, looking up with a grimace of distaste each time she did so. Then the lions moved off, side by side, leaving the crippled hyena where it lay, its tongue lolling in the sand, its death to come later.

Of all hunters, lions seem the most intolerant in their dealings with other predators. They are liable to kill without provocation or attempting to warn off any other predator they can seize hold of. Nor does this only occur in safeguarding their kills from potential scavengers. That they are prepared to go to the trouble of stalking and chasing other species of predator indicates a high degree of antipathy. Slow-moving and clumsy animals such as aardwolfs and brown hyenas are particularly vulnerable. Spotted hyenas, cheetahs and, to a lesser degree, black-backed jackals, are also occasional casualties.

Lions will, at most, only partially eat other predators, usually discarding the whole carcass. As they are not killed for food, lions obviously regard them as competitors, much as they do other lions. In fact they threaten lesser predators, not with the inexpressive facial features of the hunt, but with the bared teeth and sounds typical of their dealings with their own species. In interspecific struggles lions are also killed by other lions, but infrequently, as the adversaries are intimidated by each other's potential to inflict serious injury – a deterrent which is absent in their attitude towards smaller carnivores. To an extent, interspecific competition is reduced by

Right: *Scavenging pied crows hastily take to the air as a lioness registers her anger at their presence.*
Below: *A bloody-faced member of the nomadic Pirates looks up from a kill the two males had made. In terms of kilos of meat to effort expended on the hunt, zebra form the most important prey species for the lions of Etosha.*

different predators occupying different habitats, or using the same one at different times.

Lions are unique among the cats in that they are social animals, benefiting from the stability flowing from a communal life. Most lions live in prides consisting of males, females and cubs, and each pride resides within a fairly well-defined range that it will occupy for a year or more, and in some cases permanently. There are also nomadic individuals or groups which wander widely, following the movements of the migratory herds.

On the plains surrounding Halali, where I was based for much of my time in Etosha, the seasonal shifts of the grazers were of such magnitude that no large ungulates remained after the rains had set in. In this situation, even the two prides defending established territories were forced to abandon the plains every wet season. Initially it was uncertain what became of the prides at this time each year. It was thought they might cross into the adjoining mopane woodlands to eke out a meagre existence hunting the few sedentary kudu and hartebeest herds that lived there. Yet prides already occupied the woodlands and would in all likelihood resist any such invasion. Could it be that the prides, with young cubs to care for, would join in the hurly-burly of the nomadic life, and trail the herds to their summer pastures?

As yet, no intensive research has been done on the group structure and movements of Halali's lions, but some of the questions have been answered in the field reports of conservation officers. Through constant contacts and prolonged observation a profile emerged of all the lions using the more than 200 square kilometres of plains enclosed by the southern edge of the pan on one side, and the encircling mopane woodland on the other three.

At the heart of the plains is the Charitsaub spring which was the focus of activity of the appropriately-named Charitsaub pride, comprising three cubs aged about six months, four lionesses and two attendant males.

The Charitsaub plains are separated from the Halali plains by a wedge of mopane woodland that comes within two kilometres of the pan. The Rietfontein spring lies within this mopane belt and is heavily patronised by both plains game and woodland species. To the east of Rietfontein lie the sprawling Halali plains where the large, 16-member Halali pride had established itself. Probably because of its size, the Halali pride was far less cohesive than the Charitsaub lions and it was unusual to find all the adults together, even at a kill.

A shifting population of nomads also roamed the length and breadth of the area. The most visually dramatic grouping of nomads was four magnificent males that were sometimes in the company of two other males, the six becoming known collectively as 'the Pirates'. On one occasion, two additional males were permitted to join the Pirates at a zebra carcass and it was impressive to see the eight big males crowded flank to flank around the kill. Their manes varied in colour from tawny through gingerish to straight black. The colour differentiations were neatly paired, suggesting that the group represented four sets of siblings that had come together in this formidable alliance. The feeding males' tolerance towards each other diminished in direct proportion to the dwindling supply of meat. As tensions rose, so did the volume of rumbling growls. The two interlopers were the first to give way, discreetly withdrawing to watch the proceedings from the shade of a nearby tree. We had previously noted a blurred dominance hierarchy within the Pirates' social order, based on an awareness by each individual of the relative strength of all the others. At kills, however, none of the regular members meekly accepted a subordinate role, instead each vigorously defended his right to a share, creating a balance of power that led to a wary truce. Threats were exchanged, and although sporadic clashes occurred, little real damage was done.

Top: *Normally the best of friends, these two males clash for possession of a female in oestrus.* Left: *Two nomadic males that have spent the day resting in the shade of a spreading acacia near Salvadora spring, drink their fill in the late afternoon, before setting out for the night's hunt.* Overleaf: *The Halali flats pride sprawls at leisure in various attitudes of repose. Lions are essentially indolent and spend as much as 20 hours out of every 24 sleeping, lying about, or simply staring into the middle distance.*

Far left: *Baring his teeth, wrinkling his nose and curling his lips, a male responds with a sexual grimace on detecting that the lioness has come into oestrus.*
Left: *Emitting a harsh drawn-out yowl, the male reaches his climax. The lioness had growled almost continuously throughout mating. Though copulation is very brief, it is highly passionate and vocal.*
Below: *A young lioness lashes out angrily at her chagrined suitor. During oestrus a lioness's reaction to a male's overtures can vary from indifference, encouragement, teasing and coyness to aggressive rejection.*

The nomadic lionesses, without pride males to protect the integrity of the social unit, lived an uneasy existence. One such group that I regularly encountered consisted of a caucus of six lionesses, but as they readily associated with other females, the total on occasion swelled to nine. A single, very young cub accompanied one of the lionesses. The rest of the litter had already been lost, highlighting the vulnerability of cubs born outside the security of a composite pride.

The attitudes of the lionesses towards the Pirates depended on the personalities of the individuals concerned. One of the young females, whenever in oestrus, solicited outrageously among the males. She encouraged them all, restlessly circling the bemused object of her desire, displaying and suggestively cuffing those intimidated by the possessive posturings of her current suitor. In spite of jealous grumblings, patience was usually rewarded, as no male can maintain the copulation rate that a lioness demands for more than a few days. As a lioness will remain in oestrus for five or six days, mating on the average every 21 minutes, and do so without eating, it is not surprising that a waiting lion can expect his turn.

Most of the other lionesses avoided the males. Apart from their role in the reproductory cycle, further contacts with the males were non-beneficial to the females. It was difficult for them to raise their cubs without the secure environment of a guarded territory, and this the nomads could not provide. In fact contacts were frequently detrimental, as the males never hesitated to chase the females off a kill and, having appropriated it, refused to allow the lionesses to return until they had stripped it bare and themselves abandoned it. Strange lions are also a threat to cubs not of their own progeny and are liable to kill any they encounter if the mother is absent.

Late one afternoon at Rietfontein I observed a meeting between the six lionesses with the single cub and all six Pirates, which unequivocably demonstrated their mutual antagonism. The lionesses had spent most of the day lying up in the shade, occasionally directing cursory glances towards the churning procession of game trooping down to drink. Although they looked lean, they did not attempt a stalk, perhaps enervated by the heat. As the sun dropped and the day cooled, they drifted idly to the water's edge. After drinking they languorously groomed themselves and each other, alternatively petting then teasing the cub. They were perfectly at ease, sprawled on the short green grass, looking in the soft light like heaps of poured gold.

Then one of the lionesses jerked to attention, ears cocked, an alert and tense expression on her face, as she stared fixedly at a point on the fringe of the treeline where the Pirates had suddenly materialized. Their appearance produced an instant reaction from the lionesses. Their earlier torpor dissolved into an intense appraisal of the situation, followed almost immediately by an ignominious retreat. They bolted in a crouching trot, abdomens almost touching the ground, looking back over their shoulders in alarm. The panicked mother first seized her cub by its nape and hurried away, but, unable to make sufficient speed with her burden, dropped it and

Top: *Whirling, with bared teeth, the lioness is about to slap at the male as he prepares to dismount. A lioness remains in oestrus for five or six days during which she does not eat and expects to mate on an average of once every 21 minutes.*

Below: *Rolling sensuously on her back, a lioness basks in the glow of post-coital satisfaction.*

Top: *Two mothers supervise their cubs at a water-hole. Lionesses produce litters of one to five cubs, and because of a synchronisation of oestrus among females of the pride, several litters sometimes arrive at about the same time.*
Right: *The Salvadora pride was a marvellously integrated social unit and greeting ceremonies between members signalled their friendly intentions as well as serving to reinforce social bonds.*

turned to face the advancing males. With flattened ears she crouched, her teeth bared in a snarl, her only sound a sinister hiss. The cub gambolled around its mother, tumbling and wrestling between her front paws, unaware that its life was in danger.

As the Pirates drew nearer, the lioness seemed to contract, a tight bundle of fury, radiating a venomous hostility. The challenge was so daunting that it stopped the Pirates in their tracks. After staring intently at her for several long moments, their resolve wavered and broke and they passed her by. With bold roars they galloped out of sight in pursuit of the other lionesses disappearing into the timber.

The cub survived that exchange, but a few days later it was gone. It may have been killed by a lion or another predator, but it may as likely have been abandoned by its mother – in spite of her earlier spirited defence. Her reaction to the confrontation with the Pirates was instinctive, and her abandonment of the lone cub could have resulted from an equally fundamental set of values. Refusing to raise a single cub makes sense in terms of reproductive efficiency. If the lioness abandoned her cub, she would come into oestrus again almost immediately, and soon have another litter, with the chance of raising two or three cubs. Yet, though nomadic females probably fall pregnant and give birth as often as lionesses belonging to a resident pride, their success in rearing their offspring to adulthood is severely offset by the vagaries of their life-style.

The thronging herds and well-fed lions inhabiting the plains in the dry season reflect a situation far removed from the lush emptiness of the wet. The lions lingered until after most of the herbivores had left, as if reluctant to depart territories that had served them so well. But they were compelled to remain within range of the trekking herds. Eventually they attached themselves to the tail of the migration and disappeared along with everything else.

The Halali and Charitsaub prides left the plains just as the nomads had been forced to. We temporarily lost touch with them until six weeks later when I was able to identify the Charitsaub pride about 60 kilometres west of their old home range. The pride was still intact but appeared to have fallen on hard times; its members were ill-fed and showed it. The cubs, particularly, looked listless and emaciated.

This was our first indication that the prides followed the migration and did not merely shift their activities a few kilometres south into the woodlands. Two of the Halali males were later discovered at a water-hole not far

Below: *A cub affectionately greets its mother. The mother-cub bond ensures the safety of the pride's young.*
Right: *Two cubs from the Salvadora pride engage in a rough-and-tumble wrestling match. Young cubs wrestle a good deal during play, but as they get older, there is less tactile contact and more time is spent in stalking each other.*

from Okaukuejo and in the same general area as the Charitsaub pride had appeared. Further sightings confirmed that the members wandered with the herds throughout the wet season, although they no longer functioned as prides. The Charitsaub males split off by themselves at some stage during the summer months, and the three cubs disappeared. The large Halali pride fragmented into four different units which maintained almost no contact. Two younger males that may have been driven out were not seen again. The nomadic life proved destructive to the maintenance of pride stability and cohesion, with old associations disintegrating and new members being recruited.

Four of the Halali lionesses with their cubs returned the following dry season and re-established themselves on their old hunting grounds. Two months later they were joined by two males who had to overcome initial discouragement, but in the end successfully ingratiated themselves with the lionesses. The newcomers proved surprisingly tolerant of the cubs, to the extent of playing with them and even displaying a rough affection, tumbling them over and cuffing them gently.

The three Charitsaub lionesses also returned, but did not base themselves at the spring as exclusively as in the past. Instead they wandered widely over their old range, and although they supported themselves adequately, they had lost a good measure of their previous self-confidence. On occasions they associated with other unattached lionesses, but were either unwilling or unable to attract the sustained company of any males of whom they remained justifiably suspicious.

It was apparent that, because of fluctuations in prey availability, the plains could not support the permanent resident prides for longer than a single dry season. Furthermore, the area held by a pride one dry season was not automatically reclaimed the next. Despite appearances to the contrary, the Charitsaub and Halali prides were merely nomads that had come

together to set up a temporary territory. Neither pride constituted a permanent social unit.

Nothing prevents nomads from remaining together, and although most of their contact is transitory, it is feasible that close attachments lasting longer than the duration of a dry season could form. It appeared that, whenever possible, individuals or groups would return to a specific home range rather than just any area that supported their prey. Lionesses proved more consistent than the males, more loyal in their dealings with other pride members, and more inclined to return to familiar territory. Their maternal responsibilities probably account, in part, for their innate conservatism.

It is as hunters that lions most dramatically capture man's imagination, and hunting is more important than any other activity in the lion's day-to-day existence. Lions are great opportunists and quick to exploit a source of potential food. During the dry season the water dependence of their prey offers just such an opportunity.

At Salvadora, a strong contact spring abutting the Halali plains, the Charitsaub pride took to hunting during the daylight hours, thus coinciding with the drinking habits of the herbivores. It was a habit the lions reserved for the winter months when the cool weather diminished their natural reluctance to engage in diurnal stalks. Still, they are less effective by day and their attack/kill ratio was poor – less than a quarter of their charges succeeded – despite the fact that, to drink, the herds must inevitably enter their ambush area.

The lions' lack of success did not reduce the drama of the situation, or the thrill of expectancy experienced by those of us fortunate enough to have a grandstand view of the proceedings. From where we waited, on high ground close by the water-hole, the scene played itself out like carefully directed theatre.

One memorable day I watched the vindication of a young nomadic male whose amateurish charges had resulted in nothing more tangible than raised dust and retreating hindquarters. It was blustery, with easterly winds gusting over the open flats. In such weather the herds delay coming to drink and it was not until midday that the first animals stood poised on

One cool August morning at Salvadora spring, I watched the vindication of this young nomadic male, whose amateurish charges had hitherto resulted in nothing more tangible than raised dust and retreating hindquarters. Screened from view by a dense stand of sedge, the tautly-drawn lion waited until the right moment, then charged (Top left). *Ranging alongside the solid flank of fleeing animals, he probed for a weakness in the stampeding, dust-swirling chain of bodies – and found it* (Centre left). *A terrified zebra yearling hesitated for a moment, became narrowly separated from the rest of the herd, and in mid-stride the lion noted its distress, to change direction* (Bottom left); *intercepting it he tore the zebra to the ground and strangled it* (Top).

Clinging grimly to her kill, a lioness attempts to prevent a pride male from stealing it. When a kill is as small as this zebra foal – barely enough to fill one lion's stomach – it is every lion for itself, and the bigger, stronger males usually win any dispute over the spoils (Right).
Overleaf: *With bared teeth, slashing claws and a vicious snarl, this lioness in oestrus spurns the advances of her daunted suitor.*

the outer perimeter, carefully scrutinising the land surrounding the spring.

The lion was screened from view by a dense stand of sedge, but the zebra remained wary. They faced into the wind, shifting nervously, communicating their unease with harsh barks. The wildebeest settled to ruminate or roll with abandon in a dust wallow. They took no part in the surveillance, relying rather on the zebra's superior eyesight and caution to alert them. The tense, still lion waited and the wind buffeted the grasses. A lone heron stood motionless and guarded, keeping its own counsel.

Finally the zebras started forward, stopping frequently to re-examine the lie of the land. The lead mare reached the water's edge and stooped to drink. With fresh confidence the others surged in, wading up to their withers, eagerly gulping the tepid liquid. The lion held off, though now he had readied himself, drawn tautly together, wanting to choose the right moment. Then he made his move.

Breaking cover he streaked forward, ranging alongside the solid flank of fleeing animals at an easy gallop. I had seen him do this before and miss, and had wondered at the logic of the tactic. Now, in success, it became apparent. He was probing for a weakness in the tightly-meshed links of that stampeding, dust-swirling chain of bodies. And he found it. A terrified zebra yearling, bewildered by the sudden confusion, had hesitated for a moment and became narrowly separated from the herd. In mid-stride the lion spotted the yearling's distress, changed direction, and intercepted it. Seized in a powerful embrace, the zebra was torn off the ground. As soon as the zebra was on its side, the lion lunged for its throat and maintained a stranglehold until its prey ceased to move.

The speed and violence of a large predator's capture of its prey leaves some people appalled and outraged. Obviously, an act of such primal blood-letting is open to misinterpretation. That there is one animal less is undeniable, but there is nothing negative in the killing of one animal by another. Where life proliferates, so will death. Out on the plains, once the dust has settled and warm blood flows, an urgent, unreasoning hunger will be satisfied. There is an elementary beauty in such pragmatism. It is not ironic, but fundamental and perpetuating, that to successfully balance nature's pyramid of life, the death of a beast is required.

Cheetah at Bay

A cheetah moving at leisure exhibits the easy grace and self-assurance of a haute couture model, and it is not until it hits top speed – in a sprint that qualifies it as the fastest land mammal – that the functional design behind that loose-limbed elegance becomes apparent. A cheetah is the least typical of the big cats, its physical proportions streamlined to accentuate speed rather than power. With its small head, deep chest, trim waist and long legs, it is the embodiment of a swift, hunting feline. Its movable shoulder blades allow it to lengthen its stride by about 100 millimetres and its flexible spine is alternately shortened and lengthened in a gallop that brings its hindlegs on a parallel with its head. Twice in its sequence of running movements, the cheetah's whole body is off the ground, once with all four legs extended and once with all four bunched. With its tail acting as a rudder held straight out behind it, it can cover as much as seven metres in each stride. Its heart beats slower than an animal of comparable size – another aid in allowing it to run down prey at speeds of up to 110 kilometres an hour.

I have been in the company of cheetah on an uninterrupted savanna when they have tensed and identified as springbok what appeared to me as no more than specks on a shimmering horizon. At the same distance larger species, which did not constitute a potential meal, were given a cursory glance – then ignored. A cheetah's large amber eyes are set well forward in its head, equipping it, in common with other predators, with binocular vision. This is vital, during a high-speed chase, when even a slight miscalculation as an antelope zigzags, causes the cheetah to lose ground and, lacking stamina for a long chase, it must give up.

The first successful hunt I witnessed, dramatically demonstrated the single-mindedness of a cheetah's pursuit. I had watched the cheetah for most of the afternoon, and it, in turn, had speculatively observed a herd of springbok. The springbok had gradually grazed closer and closer to the concealed cheetah, but the cat bided its time, making no effort to close the gap. Just as the sun was dropping below the horizon, I noticed an ewe that had rejoined the herd with a new-born fawn at her side. It was the first new arrival I had seen that summer and, as the herd members crowded round to inspect it, the cheetah sighted it and immediately launched an attack.

The cheetah covered the distance at full speed, moving left to right across my line of vision, fixed and isolated by my binoculars. I felt strangely detached; it was almost like watching a moving picture as the cheetah bounded forward, bathed in the rosy afterglow of sunset, the alarm whistles of springbok the only sounds. Uninvolved springbok, realizing that they were not the object of the attack, casually moved out of the line of approach, then turned to watch the cheetah's progress. It ran hard for more than 500 metres and did not slow until it sprang on the fawn. So swift and so natural was the outcome that, thrilled by the hunt, I had no time for sadness.

During my three years in Etosha, I spent reasonably long periods watching and photographing cheetah, particularly a family comprising a female and her four cubs.

Only a few of the Etosha cheetah were approachable, the others would run away or hide on seeing a car. But the family of five had become accustomed to cars and, in time, came to regard me not so much with acceptance as indifference. The mother would lie flat on her side, her legs stretched out stiffly before her, her round head raised, scanning the horizon with an inscrutable gaze that swept by me and through me, as if I did not exist.

The guarded and aloof demeanour of the cheetah discouraged us from the familiarity of naming them. Instead they became known simply as 'the cheetah with the four cubs' or 'the Charitsaub male' as the case might be. There were so few cheetahs that there could never be confusion as to which we referred.

I first met the family of five early one morning as I drove through a tangle of *Acacia nebrownii* thorn scrub. The mother was using a fallen tree trunk as an elevated vantage point to spy out the land. I stopped well short of her and only after I had turned off the car's engine and focused my binoculars, did I notice the cubs. They were still very young, the silky, blue-grey mantle prominent from their ruffs to their rumps. The bush was too thick to be sure how many of them there were, and it was not until the mother, with an effortless spring back to earth, led them out, that I had a chance to count them. With a single, imperious, high-pitched chirp, she called them to heel, then started away with the four little bundles tumbling along behind her, wistful and vulnerable.

Thirty minutes later she sighted a lone springbok ram, busily feeding with its back to the cheetahs. With eyes fixed on the springbok, and head held parallel to her back, she hurried forward, using any available cover and freezing whenever the springbok raised its head. At 50 metres, and still unnoticed, she froze for the last time, waiting for the springbok's head to go down – and the moment it did, breaking into full stride from almost the first

Previous page: *Frightened, but unharmed by a blow struck by the female cheetah a moment before, a wildebeest calf tumbles in the dust at its mother's feet – and now it is the cheetah's turn to go on the defensive and deftly evade the lunge of the incensed cow.*
The cheetah at Etosha are extremely shy, but we were able to observe this mother and her cubs, near Namutoni, over a fairly long period.
Left: *Here one of the cubs helps its mother drag the remains of their springbok kill to the nearest shade as the other three cubs bring up the rear. At this age the cubs took a real interest in their mother's hunting behaviour, but were still unable to successfully stalk and kill.*
Below: *Looking up nervously from their feeding, two of the cheetahs scan the plains anxiously on the look-out for lions and hyenas that would appropriate their kill should they discover it.*

bound. She closed the distance to 20 metres before her quarry became aware of her rush, and by then it was too late. After a short chase, the cheetah shot out a forepaw, striking the springbok on the thigh and bowling it over in a cloud of dust. He was a sturdy old ram, however, and regained his feet almost immediately, involving the cheetah in another short chase before he was again knocked down. This time the cheetah secured a grip on his throat, and although the ram again struggled to its feet the desperate tug-of-war did not last long. As its strength failed it collapsed, the cheetah maintaining her throttle-hold until it relaxed in death.

Throughout the chase, the cubs remained at the saltbush they had dropped behind on first sighting the springbok. But now, at an urgent call from their mother, they trotted to join her. As she dragged the springbok into the shade of a spreading terminalia the cubs scrambled over the carcass, biting and worrying it, elated at their mother's success. Once the mother had settled the springbok to her satisfaction, she threw herself down under the shade, and panting heavily, rested for the next 20 minutes. She got up once to help the cubs cut through the skin of the springbok's haunches, then flopped back while they fed.

Cheetah usually abandon a kill after one sitting, but this time, probably because heavy undergrowth prevented scavengers from discovering the carcass, I found them still at it 24 hours later. The whole family was gorged with meat, their rounded bellies, tight as drums, looking incongruous on otherwise sleek frames. Of the springbok little more remained than the skeleton with most of the skin still attached, and off to one side, the discarded digestive tract.

So began my association with the cheetah family, to whose personal saga I was privy, off and on, for the next 14 months. The female was a good mother and a proficient hunter, and no mishaps intervened, so that the cubs, who I estimated to be about three months old at our first meeting, all survived the rigours of growing up, to the day, a little more than a year later, when they went their own ways as young adults.

The family spent most of that year in the vicinity of the Namutoni tourist lodge, but were liable to roam extensively. At times they were sighted at points as divergent as 40 kilometres west and 50 kilometres north-east of their base.

Several other cheetahs operated in the Namutoni area, but I saw no evidence of territoriality. Individuals and groups interchanged hunting grounds at will without any apparent friction. I never saw a meeting of strangers take place and assume, as has been suggested by ethologists, that such contacts were avoided. In the case of the family, conservation officers at Namutoni were convinced that the mother had been born in the area, and had been part of an earlier study. This would account for her tameness and possibly some of the neighbouring cheetahs were siblings, or at least close relatives.

The cheetah population in the park is surprisingly low, considering the available prey and their success in catching it. Recent studies indicate that

Top: *In a spirit of irrepressible* joie de vivre, *displaying all their characteristic loose-limbed elegance, the four cubs race across the Namutoni plains in the cool of early evening. Like all cubs, the cheetahs played every game in helter-skelter succession, developing skills and strategies that would serve them well in adult life.*
Centre: *As their mother catches her breath, the six-month-old cubs immediately start feeding on the springbok whose carcass she has dragged into the shade of a mopane after an exhausting 300 m, mid-morning chase.*
Bottom: *Yawning distractedly, the mother pays scant regard to a propositioning cub. Although she cared for them well, she remained aloof and restrained, never indulging in the uninhibited physical contact which typifies lion family life.*

Top left: *The fleeing zebra mare lags in rear-guard to her new-born foal as the five cheetahs rush forward, hoping to separate the foal from its mother's side in the confusion. The cheetahs had flopped lethargically in the deep shade of a tree adjoining the plains until potential prey moved within range; then inertia changed to concentrated tension and, following their mother's lead, the whole family charged.*
Bottom: *Provoked beyond endurance by this cheetah's probing curiosity, a frightened caracal explodes spitting from under the nose of the startled cub. Cheetah cubs take every opportunity to investigate and experiment with new 'discoveries', but the caracal had no way of knowing that the cub meant no harm.*

their relative scarcity is probably largely a result of the aggressiveness towards cheetahs – and competition with them – by other predators. Their density is far higher on cattle ranches adjoining the park, where major predators such as lions, spotted hyenas and wild dogs have been reduced or eradicated.

There have been several eye-witness accounts of the persecution of cheetah by other large carnivores. Visitors at Klein Namutoni water-hole saw a lioness catch and kill a six-month cheetah cub. The mother and her two remaining cubs were so shaken by the assault that they fled the area, and were next seen several days later more than 30 kilometres away. Accounts of lions either attacking cheetah or in the possession of their carcasses have been received from both the Kruger and the Kalahari National Parks. Miles Turner, for many years the resident ranger of Serengeti in Tanzania, found a cheetah that had been killed and stored in a tree by a leopard.

One afternoon I was attracted by a gathering of vultures and marabous at the Klein Okevi water-hole near Namutoni, and found them picking over a recently-dead young adult cheetah. It had been killed by a bite through the nape of the neck and died with a grimacing snarl on its face. The nearby remains of a springbok, and a lioness lolling in the shade of a tree, clearly suggested the course of events. As all this had taken place in the midday heat, I guessed that the lioness had been lying up and without herself being seen, watched the cheetah chase and catch the springbok. The lioness, not content with simply usurping its kill for herself, had then stalked the preoccupied cheetah and, taking it by surprise, killed it with a single bite.

That might have been the end of that, if I had not passed the same water-hole in the late afternoon. I was driving slowly, wondering if the lioness was still in the vicinity, when two furry little creatures rushed out from under a bush straight at the car. They stopped just short of me and hissed. I realized that the dead cheetah had been a mother, and these were her orphans. It was difficult to know what to do for the best. In objective terms the cubs should be left to their fate, so that nature would be interfered with as little as possible, and for the practical reason that if caught and hand-reared, past experiences had shown that the cubs could never successfully be rehabilitated back into the wilds – and zoos and such places have all the cheetahs they need.

Logic and pragmatism tend to go by the board under such distressing circumstances, however. Although too afraid to commit themselves completely, each time someone returned to the cubs they rushed up and it was hard to refuse what we regarded as a direct appeal. But the decision was taken out of our hands. Having decided to catch them we found that we could not. They were too fast to be taken on foot, and pursuit by car in such dense bush would probably have led to them being run over. We baited a trapdoor cage, but they refused to enter, although they accepted meat readily enough outside the cage. We tried for most of one day to lure them into the cage, camouflaging the bars and playing the recorded sounds of an

Consummate hunters in profile. Blood-daubed cheetahs look up from a snack of springbok fawn to stare fixedly at a point on the horizon where another fawn – another mouthful – has been spotted.

adult cheetah, without any success. At nightfall we finally gave up, and, when a search the following morning failed to find them, we assumed they had been grabbed by one of the nocturnal predators. Then two days later I saw them again, about eight kilometres from where I had last left them. It was a moonlit evening, and they rushed up as always, looking frail and without hope, yet still hissing defiantly. There was nothing I could do. I tried to edge the car closer, but they bolted into the shadows and I never saw them again. Without the protection of their mother they could not survive. The end is always saddest after such spirited perseverance.

The mother with the four cubs was careful but confident in her dealings with lions. On the one occasion I saw a lioness make a determined charge, the mother stood her ground, drawing the lioness whilst the cubs made their escape. Then obviously aware of her superior speed, she led the lioness away in the opposite direction. The cubs were nearly nine months old at the time and almost certainly could have outpaced the lioness without their mother's intervention, but never having heard of felines resorting to such manoeuvres I was a fascinated observer.

When lions had either been too far to present a threat or had not noticed them, the family stayed alert and kept an eye on them declining to hurriedly move off as I have seen other cheetah do. The mere roaring of a distant, unseen lion disturbed one cheetah so badly that, casting nervous glances over its shoulder, it slunk off in the opposite direction.

A litter of cheetah averages three or four cubs, but about half the total born die within the first few months from various causes, including disease, malnutrition and predation. One female cheetah I had come to know quite well, very efficiently raised three cubs to about six months when an injury to one of her legs affected her ability to catch prey. Although she still hunted, she had lost her edge and invariably missed. They all quickly lost condition, and tourists reported seeing the mother reluctantly abandon one cub that was too weak to follow. Soon afterwards the other two cubs were lost and, eventually, the mother disappeared as well. Without the sociability of the pride that frequently sees an injured or ailing lion through a crisis, a cheetah must rely exclusively on its own resources. I have never known, or heard of, a disabled cheetah in the wilds that did not perish.

Although they may hunt by night when the moon is full, cheetah are primarily diurnal. This greatly assisted me in maintaining contact with them, as I would invariably find them just before dawn each morning near the same place I had left them the previous evening. Around Namutoni, where game is usually plentiful, hunting cheetah only actively seek prey in the cooler hours, preferring to lie up during the heat of the day and wait for game to wander into their vicinity. Whether stalking or resting, the cheetah usually selected transitional areas between woodland and savanna, which provided cover and brought them into contact with grazing springbok.

From an early age the cubs had taken an interest in their mother's hunting activities, but not until they were seven months old was serious tuition in the art of the chase initiated. Their training coincided with the

Top: *A cub scampers out of harm's way after acting as a decoy to distract the attention of the wildebeest herd bull, and a new-born calf is left momentarily unprotected – an opportunity the female cheetah, charging in from the left, is quick to sieze.*
Top right: *The doomed wildebeest calf slowly collapses under the combined assault of the four cubs. Their mother allows them to make the kill themselves – essential practice in a technique they must perfect if they are to survive as independent adults.*
Right: *By hastily dragging this still-breathing foal into dense cover, the cheetahs prevented the rallying adult zebras from mounting a broad-fronted rescue attempt.*

springbok lambing season and the cheetahs totally ignored adult springbok in favour of the easily caught fawns. Three and sometimes four fawns were caught each day, and though I was elsewhere at the time, David and Carol Hughes filmed several instances of the mother bringing back a live fawn to her cubs, releasing it among them and then standing aside while they caught and attempted to kill it. By deliberately providing opportunities for the cubs to learn hunting skills, she was laying the groundwork on which lay their future ability to provide for themselves.

Cheetah cubs must be trained in all aspects of the hunt by their mother, for, although they are born with an inherent hunting ability, they lack the knowledge to apply instinctive responses efficiently. Cubs make such basic errors as stalking species as unsuitable as giraffe. They are over-eager, their timing poor, their judgement irrational, and even if they catch their quarry, they have no idea how to kill it. With infinite patience, using herself as a model and allowing for much trial and error, the mother invests her cubs with all the basic skills they will require in their independent future. The cutting edge of those skills is honed when they no longer have their mother to provide for them.

The easy pickings in the form of hundreds of springbok fawns is of short duration. Nearly all the fawns are born within a three-week period. Birth peaks are typical to a lesser or greater degree in most African ungulates, a phenomenon that makes young so abundant for a short period that all predators become satiated. The surviving fawns develop quickly and soon become nearly as difficult to catch as their parents.

A month after the springbok birth peak, the wildebeest and zebra began dropping their young, and the cheetahs promptly switched their attention to them. In the hunting of zebra foals and wildebeest calves, the cubs were of real assistance for the first time, as the trick here was to prise the newborn animal from its mother's side. To accomplish this the whole family attacked together and, whilst the cubs sowed confusion among the herd, the female would home in on her target. It was a thrilling sight to watch the five cheetahs rise up and stream across the plains like flashes of spotted gold, performing what they have evolved for thousands of years to do so well.

Prey species know the potential of each predator intimately, and as soon as they recognised that it was a cheetah attacking, rallied to the defence of the threatened youngster. Herd leaders vigorously counter-attacked and I once saw the female pursue a foal with a zebra stallion thundering behind

Above: *Moving at more than 100 kph and striking in mid-stride, the female cheetah seizes an adult springbok with both forepaws, hooking into its rump with her formidable dew claws. The impact unbalances the springbok and it crashes to the ground in a cloud of dust.*
Right: *Moments after the kill one of the cubs hurried over, and, imitating its mother, seized the springbok's throat in a stranglehold, worrying it, as if it were still alive.*
Bottom: *'Grinning' with fear, a black-backed jackal dodges beyond the reach of a threatening cub. The jackal had been cautiously edging nearer, in the hope of seizing a morsel.*
Overleaf: *As the cheetah family crosses their Namutoni hunting grounds, the dark of a thunderstorm gathers over roosting flamingoes on distant Fischer's Pan.*

her. Although the cheetah knocked down the foal, the stallion intervened and the little animal escaped with no more than a gash on its rump.

The cheetahs were more successful when a small family group of zebra took a short cut through a wedge of woodland where the cheetahs were resting. Although the midday temperature was well above 40 °C, the cats responded to the opportunity without hesitation. Led by the mother, they trotted towards the zebra and I lost sight of them until, suddenly, a foal burst out of the bush with the mother cheetah immediately behind it. When the cheetah struck, the foal's momentum was such that the blow knocked it head over heels. Instantly it was grabbed by the throat. Just then the rest of the zebras appeared and barking wildly galloped straight at the cheetah. They would probably have driven her off if the cubs had not come to her assistance. With little rushes forward then sudden stops, they lunged with forelegs splayed and hindquarters high, growling and spitting. The cubs' threat display balked the zebras' rescue attempt. With the cubs providing a line of defence, the mother dragged the foal back into the bush where the zebras, whose persistence remained undiminished, were unable to mount an effective broad-front charge. It was not until she was under cover that the cheetah could finally kill the foal. Earlier, distracted by the zebras' demonstration, she had released her throttle hold, each time allowing the foal to regain its breath, and once even start to rise to its feet.

The zebras seemed deeply affected by the loss of the foal, though animals are supposed to accept such inevitabilities with equanimity. For hours afterwards the herd, then just the stallion and the mother, and finally only the mother, returned to the scene, calling and calling – their only answer the moist sounds of big cats feeding.

Cheetah meals are hasty. Other large predators invariably eat the soft belly first and then consume the rest at leisure. Cheetahs, however, begin with the protein-rich muscles of the hind legs, stopping occasionally to look around, as if expecting to be interrupted at any moment. Other predators, in fact, quite frequently appropriate cheetah kills. I watched a single hyena lope over to investigate the commotion caused when the family caught a wildebeest calf. Without breaking its stride, and totally ignoring the five furious cheetahs, it commandeered the kill and settled down to eat it. There was nothing the cheetahs could do except circle at a distance, moaning with frustration.

Several months later the cubs abruptly separated from their mother, and although I did not witness their departure, I saw them several times afterwards. At first the leanness of the newly-independent cubs showed that they were having a hard time hunting for themselves, but they learnt rapidly, and managed to obtain enough to see them through until their technique improved. The two males parted company from the two females after a few months. I lost touch with the females, but saw the males once more, when they expertly caught an adult springbok. As I sat watching them eat, I thought how well they had done, how far they had come from the appealing waifs I had first known, to the rawboned hunters before me.

Elephant Trails

Elephants have come to symbolize all that is untamed and free in Africa. They have a majesty and mystery about them, a nobility and savagery that recalls the essence of the old continent. Their scream is a sound from prehistory, awakening primitive, atavistic fears in modern, civilized man.

Since the first hominid emerged in Africa two million years ago, man's destiny has been entwined with that of the elephant. Their relationship was founded on blood. The new predators, armed with sharpened bones and rocks and a superior intelligence, stalked the largest mammal on earth for its abundance of meat. They killed it and were, in turn, themselves frequently killed. In that early, fierce struggle for survival a balance was maintained – neither man's numbers nor his technology could disturb it. It was not until the arrival of Arab, then European, traders and colonialists that this balance was ruptured. In the pursuit of gold, 'white gold' (ivory) and 'black gold' (slaves), the heart of Africa was finally penetrated by the outside world. Once man had to be protected from elephants, now elephants must be protected from man.

As technological man began to settle Africa, a surge of agricultural and industrial development ultimately resulted in an unprecedented reduction in the elephant herds. Human populations that had been concentrated for safety now quickly moved to occupy land that previously had supported only wild animals. The increasing pressure of humans and their domestic stock steadily squeezed the remaining wildlife into smaller and smaller areas. Before the game disappeared completely, national parks and reserves were established, but beyond their boundaries the pressure continued to build. Centuries-old elephant migratory patterns were disrupted and harassment forced those animals living outside game parks to seek safety within.

In South Africa the demise of the elephant herds was particularly rapid. By the end of the last century, most elephants had been exterminated south of the Limpopo. In the north-eastern Transvaal, the Sabi Game Reserve, later to become the Kruger National Park, came into existence without a single elephant within its borders. However, the new sanctuary soon attracted elephant refugees from Moçambique and Zimbabwe.

In his delightful book, *Memories of a Game Ranger,* Harry Wolhuter describes his encounter with the first immigrants. 'One early morning . . . as I rode down towards the Olifants (River) I saw what I at first took to be two huge rocks away out in the sandy bed of the river. It struck me as strange that I had not previously noticed these rocks, as I had passed that way many times; and then, as I watched, one of the rocks moved distinctly and with a tremendous thrill it dawned on me that of course they were elephants.'

The year was 1903. By 1905 there were five elephants in the Letaba/Olifants area. In 1912 there were only 25 in this part of the park and none elsewhere. Before his retirement in 1946, Col. James Stevenson-Hamilton, warden of the park from its inception, estimated the number at 450. By 1960 the first aerial census yielded a tally of more than 1 000 elephants and by 1968 the number had reached a staggering 7 700. For the first time a

South African game park was faced with the question: how many elephants are too many elephants?

Worried officials came to the painful conclusion that a culling programme would have to be introduced to prevent permanent damage to the habitat. Kruger was not the only park in Africa to recognise the need for some form of man-managed control of exploding elephant populations. The alternative policy of non-interference, as Tsavo in Kenya found out to its cost, led to the devastation of vegetation and the starvation of thousands of elephants and other associated herbivores.

Etosha is no exception. In 1934 G. C. Shortridge conducted the first intensive and reasonably reliable survey to determine the distribution patterns and status of the mammals of Namibia. He found Kaokoland to be

the only area with any large numbers of elephants, which he estimated at between 600 and 1 000. On his distribution map he shows no elephants at all around Okaukuejo and along the southern edge of Etosha Pan to Namutoni. By 1954 a small group had established itself in eastern Etosha. In the west, vast areas of suitable habitat could only be used during the rainy season because of the absence of permanent surface water. But, with the development of Etosha, more than 30 boreholes have been sunk, permitting elephants to occupy northern and western Etosha throughout the year. Immigration from unprotected areas in Damaraland and Kaokoland, coupled with a healthy breeding rate, has precipitated a dramatic upsurge in Etosha's elephant population to the stage where today it exceeds 2 000.

The compression of elephants in Etosha as a result of human activities is so advanced that a familiar stage in the interaction between the two species can be recognised in the contacts across their common boundaries.

At the start of the rainy season, most elephants leave the permanent, dry-season springs which lie within the park's boundaries to wander far and wide. Before long they reach the fences that define Etosha's borders. Although these are not strong enough to contain a determined elephant's wanderlust, most elephants have learnt to resist the temptation to break through. But to some bulls such restrictions are intolerable; they trample down the fence and venture into enemy territory in their ceaseless search for food and water. Breeding herds are sometimes tempted to follow.

Initially the authorities tried to herd the escapees back using horseback commandos. Usually these succeeded, but men prepared to engage in such a dangerous game of tag were hard to come by, and the high incidence of break-outs strained the capabilities of the round-up squad.

Previous page: *His towering bulk enhanced by widespread ears, this young bull trumpets a warning. 'There is no creature among all the Beasts of the world which hath so great and ample demonstration of power and wisdom of almighty God as the Elephant' wrote Edward Topsell in the* Historie of the Foure-Footed Beasts *(1658).*

Left: *The dust of their passage swirling about them – heads up, ears back and trunks raised to test the wind – a matriarchal herd excitedly approaches the water-hole.*

Below: *Lined up at the water's edge, youngsters jostle for position between their mothers' legs, as their elders dip their trunks to slurp gratefully, then squirt water down cavernous throats with sounds like a flushing drain.*

Left: *Romping in the cool ooze churned by monumental feet, a young calf scrambles over a prone playmate. Elephants are born into a very cohesive and protective community. A close relationship forms, not only with their mothers, but with older brothers and sisters, and their boisterous play is benignly tolerated by the herd.*
Above: *Like an Old Testament leviathan, a young bull rears from the water at Agab spring, giving himself over to the exquisite pleasure of bathing. No animal enjoys itself so much at a water-hole as an elephant.*

Ultimately, the well-meant attempt to curb the elephants' footloose habits through kindness was found impractical and the tried and trusted remedy of heavy-calibre fire-power was resorted to. So an elephant that is a tourist attraction one day, goes through the fence and on to a neighbouring farm to become a 'problem animal' the next – and a carcass the day after.

Culling is always an unpleasant business – particularly so when elephants are involved. 'The great elephant', remarked Leonardo da Vinci, 'has by nature qualities which rarely occur amongst men; namely probity, prudence and a sense of justice and religious observance.' Elephants do have social customs and a range of emotional and instictive responses to which humans can relate and admire.

Out on a cull, after the stalk and the killing had taken place, the excitement died down and the adrenalin settled, the collapsed giant so undistinguished in death before us, it was hard to take much satisfaction in the day's work. I found that in the moments before the shooting started I was no longer impartial. My sense of justice was outraged. I allied myself with the elephant, silently willing it to action – either flight or attack. At least the hunters were efficient and quick.

Etosha has not reached the point where the culling of elephants is necessary within the park, but if present trends continue, this unpalatable, but unavoidable, course may have to be taken.

Left undisturbed, adult elephants maintain a quiet dignity in all they do. In the pristine spaces of Etosha there is no other animal in whose company I would rather spend long periods of time. Unlike the somnolent big cats or the dispassionate antelopes, elephants are endlessly entertaining.

They accepted my presence with an indulgent, almost condescending tolerance, occasionally ruffling my equanimity with mock-severe charges and at times completely ignoring me. I came to know one large breeding herd of 34 animals particularly well. They were led by an old, sway-backed, tuskless matriarch who, in spite of her age, was accompanied by a very young calf. She was supported by a younger one-tusked cow, probably her daughter, who also had a small calf. One Tusk was more temperamental in her dealings with cars than her mother, less sure of herself and more inclined to react with nervous threats if approached too close. A good lieutenant, her caution was really no more than a sign of the times.

The herd's movements were co-ordinated to the seasons, and consequently were as predictable. I could never be sure precisely where they would be on a given day, but I knew in which general area they would be found and which waterpoints they were currently using. At the height of winter, after frost had withered the mopane leaves in the Okaukuejo district, the herd moved east, temporarily inhabiting the extensive, frost-free mopane woodlands adjoining Rietfontein spring. As the dry season progressed, their eastward drift continued on to Nuamses, Goas and Agab until the rains came, when they headed down to the southern boundary via Dungarees. I would lose touch with them for the next six months until they reappeared again at the beginning of the dry season at Aus, far to the west.

Right: *Screaming his indignation at the temerity of so close a trespass, a young calf boldly challenges zebra and a giraffe waiting to drink.*
Far right top: *Mirror images but distinguished by their make-up, these two bulls dramatically illustrate how elephants in the wild assume the colour of the mud with which they last plastered themselves.*
Far right bottom: *After the cool wallow of a mudbath, these elephants blow dust over themselves to protect their skin from flies and the sun. Dust also acts as an abrasive helping to dislodge ticks when the elephant scratches against a tree.*

One of the surest ways of meeting up with the herd during the dry season was to wait patiently for them at one of the springs they regularly visited.

In parks where elephants are culled or have had contacts with mankind, they approach waterpoints very cautiously, and in hunting areas will usually only drink after sunset. This is certainly not the case in Etosha where, having been totally protected within the reserve for many years, they completely ignore tourist vehicles parked at the water-hole and excitedly bear down on the water at a fast walk or a run.

One hot August afternoon, as I sat reading and drowsing at Agab spring, the stillness was violently shattered by the sudden arrival of the old matriarch and her relatives. They rushed by me at a stiff-legged run, heaving grey bodies straining towards the water, paying me not the slightest attention except for One Tusk, who briefly turned to face me, ears flared, registering her protest at my presence before turning away towards the spring, casting a last angry glance back over her shoulder. Their indifference and that lone gesture of hostility was what I had come to expect and I was delighted to see them again.

Lined flank to flank along the water's edge, youngsters jostling for a position between their mothers' legs, they dipped their trunks to suck up water with grateful slurps, then squirted it down cavernous throats with a sound like a flushing drain. Very young calves with insufficiently-developed trunk muscles were obliged to kneel and gulp water directly into their mouths, occasionally slipping in the mud and squealing with fright, to be instantly comforted by the reassuring trunk of an adult.
After they had slaked their enormous thirsts (a mature elephant will drink as much as 180 litres a day) they indulged in the exquisite pleasure of bathing. Their mien changed from studied reserve to whole-hearted abandonment as they took to the water with the enthusiasm and lack of inhibitions of a gang of high-spirited teenagers. Craggy, weathered old cows dived and wallowed, snorting and blowing with hedonistic delight. Their buoyancy in water freed them to engage in stunts beyond their means on dry land. To the accompaniment of great splashes, trunks flailing the air, they tumbled over backwards, recovering only to dive head first back into the cool ooze. Calves romped after their mothers, scrambling over submerged bodies, thoroughly caught up in the excitement. Young bulls wrestled amongst themselves, engaging in trials of strength and thus establishing a dominance hierarchy that would remain into adult life. Sitting cramped and sweltering in my car, I could heartily – if only vicariously – appreciate their pleasure.

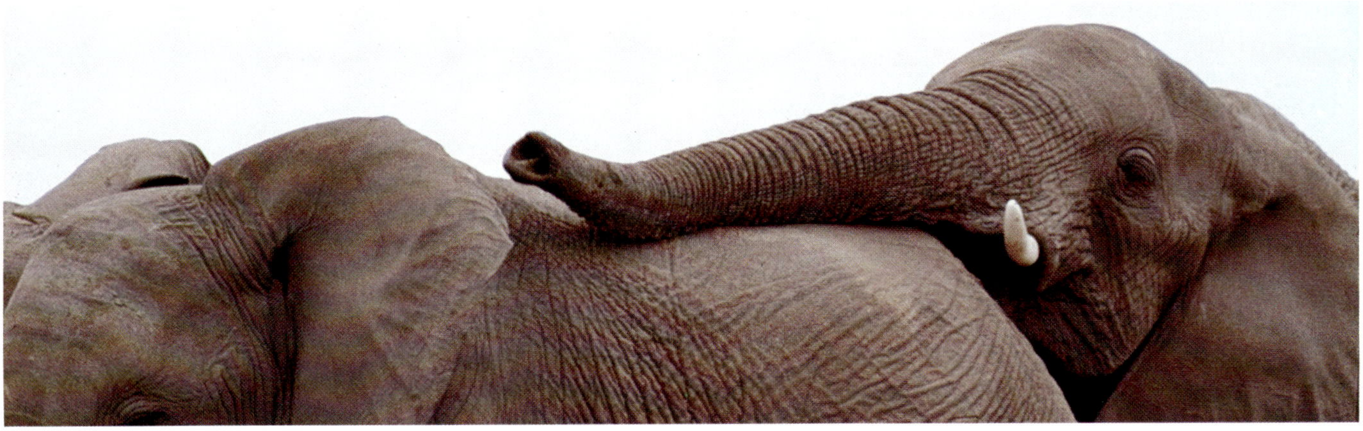

The herd spent more than an hour at their ablutions, taking no notice of me. Only after they had finished and were preparing to leave, a juvenile male decided the time had come to seek a confrontation. The prospect of facing me down obviously worried him and he signalled his nervousness and indecision by distractedly touching the temporal gland midway between his ear and eye with the tip of his trunk. For one long moment he stood facing away from me, silently brooding, perhaps 'psyching' himself up, then suddenly he turned and started towards me. He came on aggressively, but I had already decided to resist his bullying tactics and stand my ground. Elephants have a penchant for exercising their prerogative as Africa's heavyweights; but, like most wild animals, they are great bluffers and can therefore be out-bluffed; they play a game of brinkmanship, but are reluctant to commit themselves irrevocably.

Meeting at a water-hole, these young bulls communicate through a silent yet wonderfully expressive 'trunk language'. After initiating a greeting ceremony, a subordinate youngster shows his willingness to make friends. As their comradeship develops, so their gestures become more intimate.

The young bull was fully aware of the impact of his towering bulk and halted his approach long enough to stand with his forefeet on a fallen tree-trunk to add vital elevation to his authority, his threatening pose enhanced by widespread ears. When this display failed he resumed his advance, but as the distance narrowed his uncertainty showed, until, eyeball to eyeball, he played his last card. Furiously shaking his head then extending his trunk to within almost touching distance, he sounded a deep, grumbling growl, his final warning.

It did not work and I retaliated by banging the flat of my hand against the car door. The sudden, alien, metallic clanging completely unnerved him, and with a scream of alarm he turned aside, then beat a hasty, tail-arched retreat. The rest of the herd froze in their tracks on hearing my clatter, digesting and analysing the sound, then, assured that it meant them no harm, they resumed their affairs as if nothing had happened.

Elephants like it to be understood that they will tolerate no opposition from lesser animals, which are disdainfully routed – usually by the young bulls who joyfully work off excess energy by trumpeting and rushing at any antelope or zebra that dares venture too close. Even very young calves boldly challenge all strangers. This aggressive behaviour is particularly noticeable at small water-holes where nothing is permitted to drink whilst the herd is in attendance. But, if any resistance is met, an elephant is liable to back down. Authenticated tales are told of elephants giving way before the mindless courage and ferocity of an aroused honey-badger or the shrieking display of a blacksmith plover standing sentry over her clutch of eggs.

The elephant's reputation as a man-killer is largely undeserved. Their charges are mostly pure showmanship, designed to warn off rather than inflict any real damage. Yet anyone unaccustomed to dealing with wild elephants should treat them with respect. I have seen a tourist car with a hole punched through the roof by the tusk of a nursing cow. In that engagement the occupants went unhurt; but recently a policeman left his vehicle to photograph what he thought to be a single elephant, only to find himself in the middle of an incensed herd. He died of his injuries a week later.

Considering their size, elephants have a remarkable capacity for silence. Like grey ghosts – or white and even red ghosts, depending on the colour of the mud they coat themselves with – they filter through the drab bushveld, appearing suddenly, often when least expected.

A young couple, sleeping inside their camper at an unfenced campsite, first became aware of the proximity of elephants when their car began rocking gently to and fro. Looking through the window they discovered to their horror a bachelor bull relieving himself of a particularly irritating itch by rubbing against the rear fender. Early the next morning an understandably distressed woman was still hysterical when she reported the incident to the tourist office. It was suggested that the elephant was a menace and should be destroyed. The officer on duty murmured sympathetically and

Two bulls clash in a trial of strength (Below), *which will establish their respective places in a dominance hierarchy that carries through into adulthood. These struggles occur during adolescence and so prevent serious fights between mature bulls. Their sparring match ended with the stronger bull mounting his subordinate* (Left). *This behaviour is not sexual, but a non-violent way of expressing dominance.*

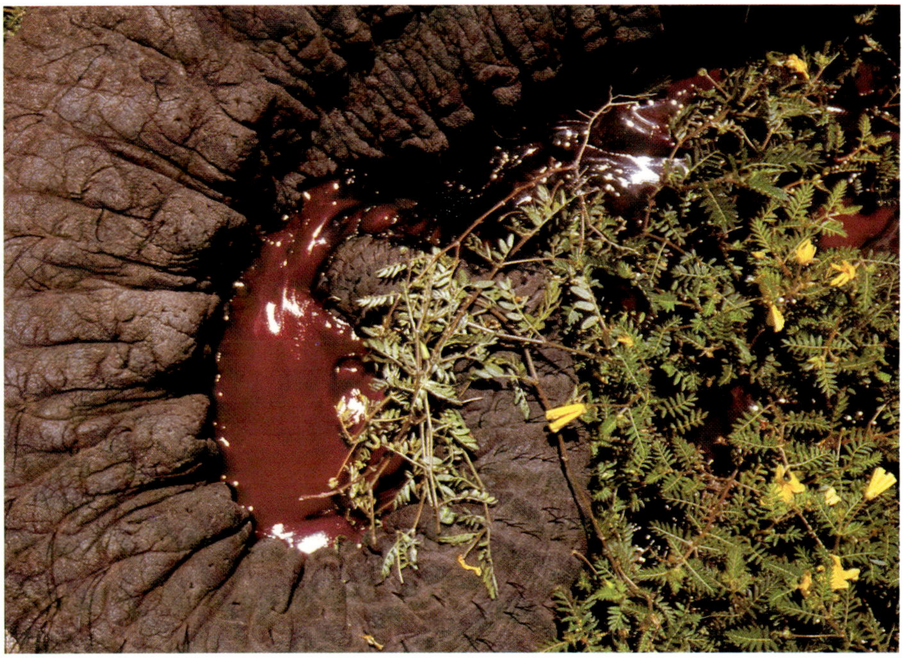

did nothing about it, but the encounter highlighted a new twist in the inconsistent relationship between man and elephants.

Where once elephants would have crashed away in panic at the approach of a vehicle, they have, over the years, grudgingly come to accept them in sanctuary areas. Today an elephant is as likely to stroll up to a vehicle as to avoid it. Conservation officials fret against the day when this familiarity might provoke an incident that ends in tragedy for a tourist. Some feel that such potentially dangerous animals must be taught to steer clear of tourist traffic. Regrettably, but probably unavoidably, the elephant's new-found trust in makind is often actively discouraged.

There are many similarities between man and elephants. Both are long-lived species, both have protracted childhoods and reach puberty at about the same age. Both live in permanent family units (although mature male elephants live apart from the breeding herds) with a highly-developed sense of responsibility towards fellow members. Both are destructive and wasteful in their methods of land use, damaging much they do not need. It is this last similarity that has caused such concern in environmental circles in recent decades – both man and elephants are running out of living space.

An adult bull elephant will spend up to 16 hours each day selecting and eating 300 kilograms of vegetable matter, and if it enjoys a full life will have processed about 4 000 tonnes before it dies. A wandering elephant's presence could hardly go unnoticed and, within their contracting habitat, the swollen populations bring unnatural pressures to bear. It is a man-made problem and man-made solutions will have to be found. Natural control factors such as starvation, disease and competition for habitat are catastrophic in a national park. Before elephants start dying of starvation or lack of water, untold damage would be done to the ecosystems which may take many years to recover. In the case of elephants, culling is the radical surgery being applied to ensure a harmonious balance. In the case of man, no solutions have as yet been found, and the future looks ominous.

Of all the African mammals, the elephant is the most difficult for man to live with. They are a measure of the dwindling wilderness areas left in Africa – a barometer that gauges the decline and destruction of ecosystems disturbingly similar to our own. If the continent can no longer support wild elephants, its impoverishment will be almost complete. The beleaguered and vanishing herds are a challenge to us to plan an environment where the needs of man will be considered and answered in relation to our entire natural heritage.

Top: *In the aftermath of a cull, the blood of a brain-shot elephant coagulates under the hot sun in silent epitaph to the death of a 'problem animal'.*
Left: *Terrifying in her fury, an elephant cow charges an intruder. Her young calf thrust under her belly out of harm's way, she leaves little doubt as to the seriousness of her threat.*
Overleaf: *Under the watchful eye of their mother, an older brother playfully kicks up a shower of muddy water, inviting his sibling to a game.*

Otjovasandu: A Wilderness

The opening up and taming of Africa was as inevitable a process as in many instances it was heroic. It manifested one of mankind's deepest, most insistent instincts, the occupation and possession of land – a drive that has left no part of the earth's surface unexplored or unclaimed. Of the vastness that was once the African wilderness all that remains today are islands surrounded by seas of cultivation and industrialization.

In an age of nitric acid rainfall and the threat of nuclear annihilation, it is tempting to indulge in a false nostalgia where the concept of wilderness is concerned. Today's sentimentalized version of a pristine paradise, without the harsh realities of unknown diseases and sudden physical danger, would have been unrecognisable to the early explorers. The European colonizing powers had no delusions about what confronted them in Africa and what measures they would need to adopt to consolidate their positions. They simply declared war on an entire continent – attacked a landmass and sought to beat Nature into submission.

'The earth', Joseph Conrad was moved to write of nineteenth-century Africa, 'was unearthly. We are accustomed to look upon the shackled form of a conquered monster, but there – there you could look at a thing monstrous and free.'

Early explorers were attracted by the freedom of Africa's vast open spaces, but to survive they had to fight the continent every inch of the way. There were wild animals, unfriendly tribes and malaria. They hunted the wild animals with an enthusiasm matched only by the naïve assumption that the herds represented a constantly renewable resource – 'peopled by wild animals alone, this region forms an inexhaustible hunting ground.'

The unfriendly tribes were neutralized by superior fire-power, but the tiny malarial anopheles mosquito was an enemy not so easily made to see reason. Thus the low-lying regions of deciduous forest, grassland and thorn bush that harboured the anopheles were avoided by settlers, becoming reservoirs for the last concentrations of wildlife. Some of these areas were later to form the nuclei of today's reserves and national parks.

In the twilight of their years, most of the great hunter-explorers of southern Africa came to deplore the extermination of the big game. One of the best known, F.C. Selous, lamented, 'during the 20 years succeeding my first arrival in South Africa in 1871, I had constantly wandered and hunted over vast areas of country, from the Cape Colony to far away north of the Zambezi, and in that time had seen game of all kinds, from the elephants, rhinoceroses and buffalos of the forest regions north of the Limpopo to the wildebeests, blesboks and springboks of the southern plains – gradually decrease, dwindle in numbers to such an extent that I thought that nowhere south of the great lakes could there be a corner of Africa left where the wild animals had not been very much thinned out, either as a result of the opening up and settlement of the country by Europeans or owing to the extensive acquisition of firearms by the native tribes.'

Wilderness areas owe nothing to the arrival of technological man and his paraphernalia. Wherever he has been, the land and its inhabitants were

Previous page: *Granite kopjes, Otjovasandu – 'The love of wilderness is more than a hunger for what is always beyond reach; it is also an expression of loyalty to the earth (the mother which bore us and sustains us), the only home we shall ever know, the only paradise we ever need . . . if only we had eyes to see.' Edward Abbey, noted American author and conservationist.*
Right: *Eyed unwinkingly by a hyrax and an agama, I rested during the heat of the day in the shade of this high granite boulder, far from technological man and his paraphernalia.*
Far right: *The pale, cascading roots of a Namaqua fig* (Ficus cordata) *squeeze into almost imperceptible crevices and fissures, eventually splitting even great boulders apart.*

diminished by the encounter. Whatever his intentions, his presence disrupted a fine balance and left it in disarray.

Some saw the disappearance of wildlife as an agreeable sign of civilization's advance. Herbivores, carnivores and insectivores (the latter accused of all sorts of unnatural crimes) were lumped together by farmers as a scourge of good farming. The wild birds and animals were regarded as either competing with, or destructive to, stock and crops. Their eradication was deemed eminently desirable. The hunters, once the writing was on the wall, redoubled their efforts to be among the last to snatch something from a declining windfall, much in the style of the bison hunters on the North American sub-continent. In no time at all the new inheritors had their monster shackled and conquered.

Only here and there has man set aside small tracts of land where wild animals survive under more or less natural conditions. In the early years, taming and exploiting the natural wealth was the task. As we have crowded the country and used up its gifts, the problems are to preserve and protect what remains.

Etosha itself is no longer a true wilderness. The very act of government required to preserve it, ensured its loss of innocence. National parks need to justify their existence. In a crowded, land-hungry world, undeveloped country must be accounted for. So tourism is encouraged, roads laid, lodges built, fences strung and boreholes sunk.

There are still places here, however, and certain times – places that can only be reached on horseback or foot. Places where wildlife is sufficiently unaccustomed to humans to resent and fear their intrusion. Where it is wise to beware of abrupt movements, and silence becomes natural.

One such place is Otjovasandu, to which I travelled as often as I could. It lies at Etosha's extreme south-western corner, where the political boundaries of Ovamboland, Kaokoland and Damaraland come together. It has never been developed nor opened to tourism. The only access road is in poor repair and holds to a straight line through featureless mopane scrub for nearly 200 kilometres. Not until Otjovasandu is reached does the reason for the journey become apparent. It is utterly different physically to the rest of the park. This is red-earth country, broken with rolling hills, dolomite ranges and granite kopjes as dramatic as Stonehenge. Squat, purple-barked sterculias push between rocks that, piled and heaped all at angles to each other, look like an unfinished explosion. Below, the country falls away, raw rock showing through open woodland with yellow patches of savanna unevenly set down. It is an African panorama – wild, rugged and beautiful.

The kopjes are remnants of the ancient rocks of Africa which have

Top left: *Otjovasandu's cliffs, dolomite ranges and boulder-strewn kopjes provide sanctuary and the high-rise lookouts a baboon troop needs to survive the dangers of savanna life.*
Bottom left: *Plucked from the water after it had over-balanced, this soaked, whimpering baby baboon is comforted by its mother.*
Below: *Duelling gemsbok bulls thrust and parry with rapier-sharp horns, capable of mortally wounding not only each other but predators. However, among themselves fights to the death are rare.*

survived the weathering of millions of years. They have stood like the ribs of sunken ships while all the softer material between them has eroded away. They serve as water catchments and, in the fissures where aeolian and eroded soils have mixed, tree seeds root that are unable to survive the harshness of the surrounding country.

My camp was not far from a wind-pump that supplied cool, pure water, filtered clean through subterranean gravel and sediment. I was close enough to collect the water on foot, but not so close as to disturb the wildlife that came to drink. The animals were timid and suspicious, even the birds would fall silent at my approach. Most had no use for mankind, only the flies welcomed me at all times. Swarms of tiny, non-biting, mopane bees danced before my eyes, attracted by moisture which they gathered to produce their own brand of honey. Their persistence was irksome and sometimes I would stop to swat as many as possible.

Water and firewood were collected each day, usually in the late afternoon. After a day in the field these were chores I associated with retreat to the shelter of my campsite. We are out of our element in the wilds at night, and a fire reassures as the sun goes down.

Sunset is the finest hour, an hour of content, with the land retreating softly into darkness while the sun fights its battle against the skyline. At peace and relaxed, listening to night come on, I sat and watched the fiery solar bowl slip below the horizon. Sometimes an unseen, restless lion roars. From the distance, faintly, its voice swells, filling all the empty spaces around it.

Top: *Straining head-to-head against each other, these two warthogs dispute possession of a mud-wallow.*
Bottom: *With a powerful kick, a mountain zebra mare in oestrus rejects the advances of the herd stallion. Otjovasandu forms the eastern limit of the rare mountain zebra's range and is as far west as the plains zebra graze. Although the two species often meet at water-holes, they ignore each other and do not interbreed.*

After sunset, the staccato calls of barking geckoes start up, exchanging gossip from rock cluster to rock cluster. About the same time, the murmured confidences of banded sandgrouse on their way to water sound overhead. The irritable bickering among a nearby troop of mountain zebra, and, once, the squeals and growls of a disaffected pair of rhino. Occasionally a spotted hyena whoops, its voice of old Africa, weird and romantic, bewailing the continent's lost wild places.

Darkness is given over to the hunters and fugitives. Frogs talk reassurance against the press of the night; small, insect-seeking bats swoop, silhouetted overhead. An eagle owl calls mysteriously, falls silent, then calls again. A trickle of primal fear touches me, then retreats before a stoked talismanic fire. Hyenas prowl, their shiftiness of the day transformed into an aura of boldness and tension. The movement of a leopard, dark shadow then silver spirit, like the flame of a fire that vanishes in the air. A wounded bird shrieks, a thin monosyllabic sound from out of the night.

I crawled from my sleeping bag early each morning, stretching to ease the cramp in my limbs. In the first sharp, cool light of dawn there is a blueness to the air and a vibrancy, shot through with the muttered chidings of countless wakening birds. Dunking my head in a bucket of cold water dispelled any remnants of sleep. I would coax the embers of the previous night's fire back into life with wood chips and kindling, adding fuel as the flames responded. Holding a mug of fresh coffee I would take my place amongst the cold-blooded geckoes and skinks on a rockface. We had come together to share the bliss as the sun's first rays massaged the chill of the night from our bones.

I woke from each night's sleep feeling completely refreshed, without any of the remorse or apprehension that has troubled me at times in other places. I could have laughed out aloud, so strong was my sense of well-being. I seemed to have at hand virtually everything I could ever need.

I spent my days roaming the country about me. If you want to know Africa, the best way is to walk it, the pulse of the place is transmitted through your boot-heel. I found the most pleasant walking along the dry riverbeds where they meandered amongst the granite cliffs. The trees, benefiting from the subterranean water, grew tall and provided a shaded, cool retreat.

Rhinos favoured these parts for their day-long siestas. Although these animals are rarely encountered, their territorial signposts – scrapes and dung middens – are everywhere. This constant affirmation of rhino occupancy added a cautionary note and tingling anticipation to my walks. Apparently-peaceful glades could never be taken for granted. The need to be constantly on guard heightened my awareness of the surroundings. The rustling of an unseen presence amongst dead leaves gained all my attention. At the cessation of a dove's cooing I would stop, equally alert. Life ·med all around – and alone and unarmed, relying on my blunted ·s for survival, I felt part of Nature's struggle.

wherever I encountered wildlife I was reminded that my presence

Far left: *A black-backed jackal basks in the first light of a winter's day. Jackals are neither mean-spirited nor moth-eaten creatures as they are often depicted, but have an alert charm and functional beauty of their own.*

Top left: *In the first sharp, cool light of dawn, there is a blueness to the air and a vibrancy, shot through with the muttered chidings of countless wakening birds.*

Bottom left: *Etched against a glowing western sky, a breeding pair of secretary birds guard their nest at the top of a tamboti tree.*

Left: *Sentinels of the kopjes. This alert pair of klipspringers were swift to bound from sight among the rocks and shrubbery of their precipitous habitat.*
Below: *Secure in its burrow, an aardwolf peers alertly at our approach. In spite of its wolf-like appearance, this rarely-encountered nocturnal animal is shy and inoffensive, sustaining itself almost exclusively on a diet of termites.*

was an intrusion. The larger antelopes fled, usually with no more than a bark of warning and a clatter of hoof on stone. Hartlaub's francolins, which live among the rocks, started up with hysterical cries at my approach. Monteiro's hornbills alternately flapped then glided across clearings, twisting their necks to look back at me. Baboons howled abuse from the safety of the cliffs. Klipspringers, perched all four feet together on a pinnacle of rock, were strangely tolerant of me, perhaps so unfamiliar with humans that they were unsure of the proper response. At the last moment they would break their fascinated stare to bound from boulder to boulder, with astonishing speed and incredible agility, until they disappeared amid the stones and shrubs far up the kopje. And once, at twilight, I disturbed a prowling leopard. It gave way without a sound, sparing me one brief glance, as cold and implacable as a winter's night, before it disappeared.

By mid-morning a welcome breeze started up, both cooling and useful in dispersing the myriad mopane bees, but as the sun climbed, the heat became oppressive.

The trees in the woodlands are mainly tall mopanes. In hot, dry weather, or at midday, their leaves, which are paired in the form of butterfly wings, fold together, so that their only shade is the shadow of their thin edges. At this time of day, with the sunlight strewn down in small spots and patches, I would rest up near the top of a kopje in the dense shade of a spreading leadwood tree.

These beautiful hardwoods, light-grey bark striped and cracked at right-angles, flourish in the alluvial soils along dry watercourses. They are called Omumborombonga by the Hereroes who revere them as their ancestral tree, from which came human beings, their flocks and herds and all the wild

beasts that exist today. They still sometimes acknowledge it with filial respect – 'I greet thee, O my father' is the appropriate salutation.

Amongst the sprawl of granite boulders, stone-eyed agama lizards sunned themselves on the rock. The females have bright green patterned heads, but are shy and reluctant to venture too close. The males are a brilliant blue and orange. Truculantly, they demonstrated the quick press-ups of agitation thought to be territorial threat displays. I established good relations with one of the males by feeding him a steady supply of flies that had pestered me. He caught on quickly and would scuttle forward with keen anticipation as I manoeuvred to strike whenever a fly settled. Before long he had taken up a position on my knee, watching my hand with unblinking encouragement, then snapping the stunned fly straight from my fingers. We were a formidable combination and mutually appreciative of the benefits our partnership afforded us. I was no less grateful for the trust he extended than he was for the supplement to his diet. After he had eaten his fill he waddled off, leaving me to wonder at the perfection of our meeting.

Bright-eyed hyraxes gathered along the uppermost ledges of the kopjes to peer down at me with barely-suppressed alarm. But I found that they, like most creatures who have had few dealings with predatory man, soon took my presence for granted. Providing I made no sudden movements, they went about their business in their desultory fashion, unconcerned by the foreigner among them. Their real danger lay in the unexpected visits of

Far left: *Lunging at his opponent's throat, a rearing stallion bites a mouth-hold of tough skin. Though they can inflict vicious-looking wounds, fighting zebras rarely do serious harm.*
Top: *This orange and blue male* Agama planiceps *spent much of his time chasing and bullying a vividly patterned female of the same species amongst the sprawl of granite boulders.*
Left: *Hood spread and scales glinting in the sun, a Cape cobra raises its head with an angry hiss.*

resident black eagles. A pair nested in the top of a tall *Acacia albida*, as had their forebears for many generations, preying on the predecessors of the valley's present hyraxes. Sweeping close by the jagged sides of the kopjes, the eagles relied on sudden appearance to seize any unsuspecting hyrax.

Ghosts inhabit these valleys and hills. The rush of wind through the eagle's wings has started them whispering. They are the spirits of the Early Race – Bushmen trackers softly conferring. They are abroad at this time in the heat of the day, because as all Africans know, this is the hour of the walking dead.

Their Stone Age artifacts are sometimes found, and the air rings with their presence. But the people are gone. Their hunting lifestyle can no longer be tolerated. It is obsolete, and they must learn to come to terms with a different civilization in some poor shanty town. Here they have traded their dignity and independence for alcohol and despair.

A race of Bushmen, known as the Heikum – which means 'people who sleep in the bush' – once had sole possession of all these lands. They lived a hunter-gatherer existence in clans and loosely-connected family groups without chiefs or a tribal entity. They possessed a remarkable ability for living off the land, subsisting on what the earth provided in a basic form. Their extremely hard life offered few choices – only necessity. They knew where water could be found, recognised the trees and bushes that gave edible fruit and where to dig for wild tubers. They were an integral part of the ecosystem in which they lived.

In the past two centuries, these little nomads were quickly crushed between the European hammer and the Bantu anvil. They were not 'dispossessed' for they recognised no land ownership, either amongst themselves or anyone else. They were as much a part of nature as the wild animals they hunted and, just as thoroughly as the wild animals, they themselves were driven off or killed.

The Bushmen, like all aboriginals, have been degraded through contact with modern civilization. Their social fabric and discipline were so finely balanced that they fractured and collapsed at the first pressures. With grieving, grinning, dissolute faces they watch the passing of the old ways, not pretending to understand the new. They have come to believe in the scorn others direct at them and parody themselves as scoundrel and fool.

Perhaps the Bushman is just another anachronism whose passing is a regrettable, but inevitable, side-effect of 'our manifest destiny'. But where will it end?

Viewed from the heights of the kopje, the country below me stretches away, ageless and remote. Earlier white men who had loved the wilderness referred to it with gruff affection as MMBA – miles and miles of bloody Africa. It must have seemed so paramount and enduring then.

In the end I believe that I too will have no place left in Africa. But I shall stay as long as I can, for Africa has marked me and I shall never be the same again. In the time that remains, I have my camera and my curiosity, to investigate all and everything, as much as I please . . .

Left: *Of the honeybadger Laurens van der Post wrote 'I know of no creature in the world so without fear, so dedicated to its own way of life and with so much of the magic of the beginning clinging to its spirit'.*
Below: *In an undulating, rocking-horse motion which keeps hindlegs and forelegs together, giraffe thunder past my camp site, their gallop quite unlike their walking gait in which the legs on each side move simultaneously.*
Overleaf: *A herd of elephants drifts across the plains and into the coming night.*
Last page: *As the full moon stares in silence at the earth, a heraldic cheetah keeps watch.*